Blue Oasis No More:

Why We're Not Going to "Beat" Global Warming
and
What We Need to Do About It

Glenn A. Ducat, PhD

DEDICATION

"Feeling gratitude and not expressing it
is like wrapping a present
and not giving it."

Writer William Arthur Ward, 1921 - 1994

This book is dedicated to my parents, Elizabeth O'Delia Baumann Ducat and Alexander Chalmers Robb Ducat. Without them, I would (literally) not be here.

While my parents might never have understood the technology I pursued in graduate school, they were, nevertheless, proud that I earned a PhD. Their unwavering commitment to and belief in the value of education were a big part of what propelled me through school. I think the same could be said for my sister, Diane, who also earned her PhD and taught at a college.

I'm also dedicating this book to my primary thesis advisor at MIT, Michael Driscoll. Dr. Driscoll was the most inventive, hardest-working, down-to-earth, "idea generator" I've ever known. I can't tell you how many times I showed up in my graduate student office only to find a multi-page, hand-written note from him, complete with back-of-the-envelope calculations and the admonition: "Have you thought about this?" Years later I learned that virtually all of his advisees regularly received "Driscoll-grams." He is the reason I, and literally hundreds of his advisees, "thought" about a lot of things.

MIT is where I learned how to think about problems. I can't know if Dr. Driscoll would have agreed or disagreed with the conclusions I reach in this book, but I hope that my thought process would have stood up to his scrutiny.

ACKNOWLEDGMENTS

**"I get by with a little help
from my friends."**

**Song Lyrics,
John Lennon and Paul McCartney 1967**

I greatly appreciate the assistance of a number of individuals who helped me refine many drafts of this book. Among them are my sister, Dr. Diane Ducat, who helped me remove many of the personal indulgences that crept into early drafts and focus on you, the reader, and my central message. Many thanks also to personal friends, Mike Buhbe, Bob Funari and Bob Scheffler for their input and feedback. Special thanks to my friend and fellow Cornellian, Judge Brett Klein for meticulously reading what I mistakenly thought was a near-final draft. His eye for detail made this a better book and me a better writer. Thanks also to Ben Preston at the Rand Corporation for comments on an early draft. They all helped identify gaps in my argument or passages that were unnecessarily long, repetitive or confusing. Thanks to Jim Nations who provided advice about self-publishing and Patricia Terry for capturing the photo that appears on the back cover..

And finally, I would like to acknowledge in advance the contributions of what I hope will be a number of scientists, researchers, and other interested parties whom I openly solicit in this first edition to provide input for the 2nd edition. Their input, including comments, corrections, points of agreement <u>and</u> disagreement on a range of issues will contribute to what I hope will be a much-improved 2nd edition.

GAD, 2022

TABLE OF CONTENTS

"Just Try To Make It Sound Like You Wrote It That Way On Purpose."

Bill Murray, *The French Dispatch*, 2021

Heat Balance 101

Energy: Where It Comes From and Goes To

TABLE OF CONTENTS (continued)

Why This Is Such a Hard Nut to Crack

Looking Ahead

LIST OF EXHIBITS

"A picture is worth a thousand words."

Ad executive, Fred R. Barnard, 1921

LIST OF EXHIBITS (continued)

EPIGRAPH

"We have met the enemy and he is us."

Pogo
Cartoonist Walt Kelly, 1970

A *POSTHUMOUS* FOREWORD
"*BY*" RACHEL CARSON

"There is nothing more difficult to take in hand, more perilous
to conduct, or more uncertain in its success, than to take
the lead in the introduction of a new order of things.
Because the innovator has for enemies
all those who have done well under the
old conditions, and lukewarm
defenders in those who may
do well under the new."

Machiavelli, The Prince, 1513

People have portrayed me in plays and written about me extensively, but as far as I know, this is the first time my name has ever been invoked as having "written" a **posthumous** *foreword to any book. I'm flattered . . . I guess.*

I am, of course, dead, and have been so since April 14, 1964. So, it's patently absurd for the author of this book to include a foreword "by" me, Rachel Carson. I can only assume that the author is attempting to generate a bit of controversy with an eye toward selling more books. But then again, he's pledging to donate all royalties to environmental and humanitarian causes, so maybe he's just a fan.

Nevertheless, "controversy" is not a bad strategy. Controversy SELLS. While I never set out to be controversial myself, I experienced more than my share of it, even in the comparatively less-polarized days of the 1960s. Less polarized, perhaps, but no less fierce.

From an early age, I was drawn to writing. That was my passion. But my mother also inspired in me a love of the natural world. My first serious writings were devoted to sharing the wonderment of the open oceans with my readers. I wrote three books about the oceans: The Sea Around Us, The Edge of the Sea and

Under the Sea Wind. _The trilogy was well received and a joy to share with so many readers at a time when "science" was not often found on best-seller lists._

_Most people, if they remember me at all, probably associate my name with my last book, _The Silent Spring_, published in 1962. That book was more than ten years in the making and dealt with the indiscriminate use of chemical pesticides. I first became concerned about the side effects of these chemicals in the mid-1940s, but that concern took a back seat to the practical obligations of caring for my mother and earning a living as a biologist and communications specialist. I was only able to return to the issue of pesticides in the environment a decade later._

A variety of reports documenting the environmental impacts of chemical pesticides began appearing in the 1950s. I was familiar with those kinds of dense, highly technical reports from the writing I had done during my years with the Fish and Wildlife Service. One of the primary chemical culprits identified in those reports was a product manufactured by DuPont called dichlorodiphenyltrichloroethane, DDT.

_DDT had been hailed as a modern "wonder" chemical. It had helped the Allies win World War II by wiping out the mosquitoes that were spreading malaria, a disease that had reached epidemic proportions among troops fighting in tropical war zones. The discovery of DDT even earned Swiss chemist, Paul Müller, a Nobel Prize in Medicine in 1948. By the late 1940s and early 1950s the widespread, indiscriminate use of DDT was wreaking havoc in the natural environment as concentrations of the chemical built up in the food chain. Massive numbers of birds were dying, birds whose songs had been silenced – hence the name of my book, _The **Silent** Spring_._

My book included a summary of published and unpublished technical reports documenting the enduring damage being done to the environment by synthetic chemical pesticides. Contrary to what many may have thought, I wasn't anti-chemicals, or even anti-pesticides. I was, however, fervently opposed to their indiscriminate use and their use as a first choice, rather than as a last resort.

As a scientist, I may have had the naïve expectation that pesticide manufacturers would revise their protocols for the use of the chemical after reviewing the data I had assembled. Instead, the U.S. Department of Agriculture and the chemical industry seemed to be incensed that anyone would question anything they were doing. After all, DuPont's marketing slogan at the time was "Better Things for Better Living . . . Through Chemistry."

In those days government was generally trusted and the actions of corporations went largely unchallenged. By exploring what today might be called an "inconvenient truth," I knew my book, was going to ruffle some powerful feathers. Still, I was unprepared for the vehemence of their response.

Scientists make observations, generate hypotheses, perform controlled experiments and record the results. They then compare their observations with what their hypotheses predicted and report both, along with a description of what they did in sufficient detail to allow their work to be replicated and checked independently. Their findings are available for the rest of the scientific community (and the general public when appropriate) to review, challenge, use, extend or . . . ignore.

The Silent Spring wasn't ignored. Perhaps my critics had never heard the expression, "Don't shoot the messenger." But shoot they did. They developed a coordinated campaign to discredit my message in much the same way the tobacco industry later responded to its critics and the fossil-fuel industry has responded to global warming concerns. In this case, they also came after me personally.

I was attacked because I was a **woman** *-- as if to imply that a woman couldn't possibly have the intellectual capacity to do "science." I was also attacked because I* **only** *had a Master of Science degree, as if to imply that only PhDs could gather, organize, evaluate and present data of significance. This was the early 1960s. Can you imagine the onslaught to which I would have been subjected in the internet era with Facebook, YouTube, Twitter and all the rest?*

It should be noted that none of the attacks were directed at the data; the data were unassailable . . . and they knew it.

Most scientific tomes (many of them important) never reach a wide audience. Indeed, The Silent Spring might have slipped into obscurity had the response of the chemical industry not been as strident. Ironically, I may have them to thank for catapulting my book to the top of the bestseller list and inextricably linking my name to the emergence of the modern environmental movement.

In the 1950s and '60s, bird populations plummeted. The birds were dropping dead or failing to survive incubation in the nest due to thinned shell. Both causes of death were consequences of the DDT we were putting into the environment. The silence of the birds was deafening. They were sending us an SOS. We just had to listen.

Today, nature is again sending us an SOS as a result of what we are now dumping into the environment. The lesson of the 1950s and '60s was that the only truly sustainable path into the future was to live in balance with nature, a harmonious partnership. We didn't learn that lesson then and mankind has again veered recklessly off that harmonious path. This time we're not just out of balance with nature, we're catastrophically out of balance.

The issue I was dealing with in The Silent Spring was the indiscriminate use of chemical pesticides, DDT in particular. In that book, I wrote these words:

> "We stand now where two roads diverge. But unlike the roads in Robert Frost's familiar poem, they are not equally fair. The road we have long been traveling is deceptively easy, a smooth superhighway on which we progress with great speed, but at its end lies disaster. The other fork of the road – the one less travelled by -- offers our last and only, chance to reach a destination that assures
> **THE PRESERVATION OF THE EARTH.**"
>
> *Rachel Carson, 1962*

(Emphasis added)

The threat today isn't chemical pesticide; it's global warming. If I were alive today, I would most assuredly be a climate activist. I would have tried to remain optimistic, but the earth today is so far out of balance and the remedies so far out of reach, I would probably have had to agree with the author's central conclusion that we're not going to "beat" global warming.

Knowing what we now know about global warming and its likely consequences, we can either try to reestablish a harmonious partnership with nature or we can continue on the superhighway to an unpleasant destination. It's our choice, but our children and their children will ultimately answer the question, "Did they do what they needed to do to 'beat' global warming?"

PREFACE

**[If you want to be remembered]
"Write something worth reading or
do something worth writing [about]."**

**Benjamin Franklin,
<u>Poor Richard's Almanac</u>, 1738**

This book has been rattling around in my brain for fifteen years. In that time, much has changed in regard to global warming and climate change, but my conclusions have not. Now that the book has become a reality, I hope that it adds to the global warming discussion and motivates a reconsideration of what we need to do to lessen the impacts of global warming on us and future generations.

I wish someone else had stepped up to write this book, someone well-respected enough or at least well-known enough that their words might carry a lot more weight than mine. But that didn't happen.

When I first started thinking about how a book like this might contribute to the conversation, I viewed global warming as an issue of **inter-generational justice:** "We have a

> *Combating climate change is not only an issue of inter-generational justice, it's also a quality-of-life and survival imperative.*

responsibility to fix this for future generations." In the intervening decade+, the time frame for the onset of the impacts of global warming has been brought dramatically forward. Those once-distant impacts are being felt today. Global warming is not only **an issue of intergenerational justice, it's also a quality-of-life and survival imperative**. Climate change caused by global warming is here NOW.

Lots of prominent authors have written about CO_2 and climate change. I've read several of them, certainly not all. In 1989, Bill McKibben published The End of Nature, perhaps the first widely read book on climate change. In 2006, Davis Guggenheim's documentary film *An Inconvenient Truth* was released. In it, former Vice President Al Gore gave his now-famous global warming awareness lecture.

I applaud what these and many other authors and filmmakers have done to raise awareness of global warming and promote policy changes to combat it. My primary complaint, however, is that most of their solutions are either superficial, too optimistic, or inadequate to materially change the global warming trajectory we're on.

As important as the 2006 Guggenheim/Gore documentary was in raising public awareness, I'm afraid the underlying upbeat, "we-can-beat-this-thing" message was wrong. The reasoning seemed to be: "If we could send men to

> *Most authors addressing global warming offer solutions that are too optimistic, unrealistic, or inadequate to alter the climate change trajectory we're on before it's too late.*

the moon, surely we can fix this." The documentary's creators either knew the sad truth and couldn't bring themselves to reveal it or they hadn't thought deeply enough about the issue to warrant their optimism. **In 2006 it was already 50 years too late to "beat" global warming, not because of the CO_2 burden the atmosphere was carrying at the time, but in terms of the dynamics of change.**

To the best of my knowledge no one else has written a book with this message. I'm certainly not the only one who has come to this realization, but I may be the first one who's been willing to put it in writing and explain why, in some detail, we're not going to "get 'er done." Maybe nobody else was willing to be the bearer of bad news. Most other writers on the subject seem to be content to highlight the severity of the issue and the direness of the situation but have somehow convinced themselves that we still have a chance to handle this. Maybe they thought that if they didn't offer hope, they might discourage meaningful movement in the right direction. If so, I think they were wrong. I'm afraid the overly optimistic message, ("we CAN beat this") has contributed to the complacency that persists today. That was certainly **not** the intent of those authors, but it may have been the subliminal takeaway.

The conclusion I reach **DOES NOT** mean we **SHOULDN'T** do everything we can do to crash down our CO_2 emissions. For our own wellbeing and for future generations, we have to. In fact, we WILL do a lot toward that end over the next several decades, but we should be under no illusion that we're going to "beat" global warming. Accepting this simple reality is crucially important in terms of deciding how to move forward. It's only after we fully understand that we're not going to "beat" global warming that we might be open to the full array of things we need to do. We're not going to "beat" global warming. That part of my message is not uplifting, I know. But that's only half the message I'm delivering. The second half, "What We Need to Do About It" is no less important. If there was nothing could to do, this would be quite a different book. But there are things we need to do, quickly and convincingly.

If you read this book and feel despair, I'm sorry. I can't control how you might react to the facts and arguments I lay out. Despair is not what I feel. I'm enough of an optimist to have written this book, which I hope will motivate more people to demand much more action. In these troubled times, I would give anything to be releasing a book with an

> *Unfortunately, we should be under no illusion that what we'll be able to accomplish over the next few decades will allow us to "beat" global warming.*

uplifting and hopeful message on almost any topic. This isn't one of those books. While the message I'm delivering will be disconcerting to anyone who thought we could actually "beat" global warming, I hope this sober assessment and view into the future will equip all interested parties with a perspective that will make them more realistic, more effective, and more insistent that we act boldly and quickly.

In reality, there's no good time to release a book with the underlying message I'm delivering. Nevertheless, I remain hopeful that this book will open some eyes to what really needs to be done to transform the world energy economy away from fossil fuels . . . and what a substantial challenge that will be. Hopefully, the content of this book will release the creative energies of others who are even more immersed in this issue or who have specialized expertise to come up with something that can at least defer the worst consequences of global warming further into the future.

PROLOGUE

**"If you are going through hell,
keep going."**

Winston Churchill

Imagine you're the coach of a high school basketball team. It's the homecoming game. It's halftime. Your team is losing by a score of 54 – 17. It would be an understatement to say it's not going well.

On the way back to the locker room, your star player twisted an ankle. Half your team's fans have left to get pizza and the other half is still in the stands, but they're otherwise absorbed on their phones. Short of divine intervention, it's pretty clear that the game's not going to end in a glorious comeback victory. And besides, you know that banking on divine intervention has never been a high-percentage strategy for success.

The door to the locker room closes. You address the team.

"OK team, I know things look pretty grim, but . . ."

1 EXECUTIVE SUMMARY

"The facts, ma'am, just the facts."

**Misquoted, but popularly attributed to
Sgt. Joe Friday on TV's Dragnet**

You're probably not going to like the conclusions in this book any more than I do, but I think they comprise a perspective that needs to be heard and understood. All I ask is this: "Please don't shoot the messenger."

We're not going to "beat" global warming.

By that I mean, we will not reduce the concentration of CO_2 in the atmosphere quickly enough to prevent the continued heating of the planet and the resulting climate change that will completely upend life as we've known it.

Global warming is real and is caused by human activity. Increasing atmospheric concentrations of CO_2 from the burning of fossil fuels and the buildup of other greenhouse gases have increased the fraction of the sun's energy being absorbed by the earth. If we don't stop this excess heating by first reducing our CO_2 emissions to zero and then reducing the excess CO_2 we've added to the atmosphere, the earth will continue to heat up until it settles in at a higher equilibrium temperature consistent with the concentration of greenhouse gases in the atmosphere at the time.

The first step in trying to avoid catastrophic changes to the earth's climate caused by human activity would require nothing less than a complete transition of the entire world's energy economy to carbon-free energy sources in less than three decades – a monumental and unachievable undertaking. That first step, while insufficient to "beat" global warming, will at least stop the problem from getting worse.

We need to stop burning fossil fuels and electrify virtually everything we do with carbon-free energy resources. This means "decarbonizing" not only the existing electrical grid, but also displacing the much larger quantities of fossil fuels used in transportation, residential, commercial, and industrial activities.

Nuclear power is a "carbon-free" technology that could help reduce fossil-fuel emissions, but in the U.S., public opposition to nuclear power coupled with the perception that solar and wind power will "solve" the global warming problem, will delay the widespread resurgence of nuclear power until 2050 or 2075 - well after it's too late to avert catastrophic environmental damage.

Without nuclear power, wind turbines and solar photovoltaics are the only practical, widely deployable, carbon-free energy technologies to replace the burning of fossil fuels. All other major carbon-free power generation technologies are either too site-specific, already fully exploited (or nearly so), unproven, unsupported by the general public, more expensive or have significant environmental issues of their own.

The problem is, both wind and solar resources are intermittent in nature. They are available only when the sun is shining or the wind is blowing. Therefore, in order to maintain the same level of electric system reliability that we currently enjoy, repowering the entire energy economy with intermittent energy resources will require massive investments in bulk energy storage capacity **and** the additional carbon-free generating resources whose sole function will be to charge that storage capacity. In addition to this added infrastructure, we'll have to greatly expand our high voltage transmission lines to move energy from where it's generated to where it's consumed. In the absence of adequate investments in this added infrastructure, utilities will be forced to run their fossil-fuel power plants at night and at other times when the renewable resources are unable to satisfy demand. The net result is that squeezing fossil fuels out of the existing electrical grid will be expensive, slow, controversial and, at best, only partially successful.

Electrifying the transportation sector, currently running primarily on petroleum products, and replacing the direct combustion of fossil fuels in the residential, commercial, and industrial sectors with electric alternatives will be an entirely **new** electrical load. The electrical grid needed to support that transformation will be at least three or four times the size of our current grid.

Displacing all the fossil-fuel consumption in just the U.S. with carbon-free technologies would require the deployment of:

20.7 billion (residential size) solar panels
OR
1.3 million medium-size wind turbines
OR
1,180 large nuclear power plants.

An undertaking of this magnitude (if it were even possible) would take many decades to complete, decades of time we don't have. As daunting as these decarbonization numbers are, they'd actually have to be at least 20% greater to accommodate population growth and replace aging carbon-free resources over the almost-thirty-year period between now and 2050.

Substantial quantities of carbon-free generating resources **will** be deployed in the U.S. and around the world, but only a fraction of what will be needed to completely eliminate CO_2 emissions. An even smaller fraction of the necessary storage capacity will also be deployed. As carbon-free resources and storage capacity become a larger share of supplying the world's energy mix, energy costs will rise substantially. As they rise, public support for decarbonization efforts will wane. Vested interests will exploit public disenchantment and attempt to derail any decarbonization changes that might disadvantage their enterprises.

International efforts to achieve a sustained response to reduce CO_2 emissions will be contentious because not all countries will be equally impacted by a warming climate or by phasing out fossil fuels. It will also be contentious because not all countries are equally responsible for the buildup of excess CO_2 in the atmosphere. International efforts to develop, implement and enforce an agreement to combat global warming will fall well short of what would ultimately need to be done.

At the end of his book on climate change, Bill Gates declares, "I'm an optimist because I know what technology can accomplish and because I know what people can accomplish." He continues: ". . .We can avoid a disaster . . .We can preserve the planet for generations to come."

I too am an optimist, but I'm also a realist. I know that technology has limits. I also know that change takes time, that vested interests seldom look beyond their own bottom line and that people are . . . you know, people, who, along with all their creativity and energy are also endowed with frailties and flaws. Mr. Gates may be half right. Perhaps we could avoid a climate disaster, but we won't. And, even if we could somehow garner the necessary world support to do what we would need to do, those actions would be so disruptive, the "cure" would introduce its own set of painful consequences.

Earth, our little blue oasis in the vast darkness of space, literally billions of years in the making, is undergoing a catastrophic transformation in the geological blink of an eye. Climate scientists tell us we need to limit the earth's global average temperature rise to less than 1.5° C. We're likely to blow through that "maximum allowable" temperature increase limit in the late 2030s and twice that amount before the end of the 2070s – and that's assuming we take some aggressive actions to reduce CO_2 emissions.

Our era, the fossil-fuel era, will ultimately be judged to have been reckless and irresponsible. The world we leave our grandchildren and great grandchildren will be immeasurably less hospitable than the one we inherited. This will not be the end of all mankind, but it will redraw world maps, claim countless lives, set in motion mass migrations and upend life as we've known it. If we do everything we can and we're wildly successful, we may be able to delay the onset of the worst consequences of climate change, but at present, we're not even on that less ambitious path to meaningfully lessen the magnitude of those consequences.

We will NOT "beat" global warming.

2 INTRODUCTION

"We are the first generation to feel the sting of climate change. And we're the last [one that] can do something about it."

Washington Governor, Jay Inslee, 2014

This book poses and answers the question:

Will we "beat" global warming?

I'll be using the terms "global warming" and "climate change" almost interchangeably though they are different. Global warming relates to the absorption by the earth of excess solar radiation resulting in one kind of climate change: persistently higher global average temperatures. Absent any other outside influences such as massive volcanic eruptions, this warming will drive other changes in long-term average regional weather conditions, i.e., climate, such as rainfall, humidity, cloud cover, average wind speed, wind direction, etc. . . . and temperature.

Unfortunately, the terms "global warming" and "climate change" don't convey the sense of an impending disaster. The term "global warming" may even conjure up an image of being wrapped in a warm blanket on a winter's night. That's not what this is.

Global warming is an environmental crisis with the potential to do immeasurable harm to our planet. I will use the terms "global warming" and "climate change" because they're part of the lexicon, but I'll also call it what it really is: a CGI, a **Catastrophic Global Imbalance** (not to be confused with the Clinton Global Initiative). By the end of this book, I hope that every time you hear or read "global warming" or "climate change," you'll understand that we're really talking about a **Catastrophic Global Imbalance.**

Clarifying The Question

Other than being concise, the question I've posed ("Will we 'beat' global warming?") is a lousy question. It's imprecise. What would it mean to "beat" global warming?

There are two ways climate scientists try to help the public understand what needs to be done to combat global warming. One approach is to explain the need to limit the earth's warming to no more than 1.5° C or 2.0° C. The other is to specify a maximum concentration of CO_2 (parts per million) in the atmosphere that should not be exceeded.

Unfortunately, the general public can't relate to either of these quantitative measures. They can't see "parts per million" and while most people sense that the earth is warming, their attention is more focused on and affected by today's weather and local temperature rather than climate and the earth's temperature. At best, the public understands that staying below these limits is better than going above them, but by-in-large, it doesn't understand **what it will take to achieve these criteria** and **what will happen if we don't**.

For the purposes of this book, the question: "Will we 'beat' global warming," is shorthand for:

Will we (humankind)
reduce our greenhouse gas emissions and
their concentration in our atmosphere
quickly enough to avoid
major, recurrent, weather- and climate-related
economic losses, human losses
and ecological damage
so that life,
as we've known it,
can flourish in the future?

This much wordier version of the question is what I mean when I pose the shorthand question: Will we "beat" global warming?

Spoiler Alert: the answer is "NO."

In fairness, I need to pause here to pose an entirely different kind of question.

If there were a genetic test to determine if you were at risk of inheriting a serious medical condition, would you take the test? Would you want to know?

If your answer is "No, I wouldn't take the test," maybe you shouldn't keep reading this book. Close it now, pass it to a friend and continue on in "blissful unknowing." If, on the other hand, you would take the genetic test to understand more fully how your future might unfold, keep reading.

Just as it was necessary to clarify the global warming question, I also need to clarify the answer.

Global warming is not a binary condition. It's not like we'll have no global warming if we keep the atmospheric concentration of CO_2 below some magic number and apocalypse if we don't. **Global warming is a continuum.** That's both the good news and the bad news. The bad news is that global warming is happening now. (That's why I'm using quotation marks around the word "beat.") The earth is warming now and will continue to warm as long as more energy is being absorbed inside our atmospheric envelope than is being radiated back into space.

The good news is that if we reduce our CO_2 emissions, we can slow the rate at which we're making things worse. In the recent past, however, the concentration of CO_2 in the atmosphere has not only not gone down, it's been going up. And, it's been going up at an increasing rate. The earth's energy imbalance is getting larger with each passing day. When I answer the question this book poses by saying we're not going to "beat" global warming, I'm saying **we won't reduce atmospheric concentrations of CO_2 enough, nor will we reduce CO_2 concentrations quickly enough** to stop global average temperatures from increasing to a point that completely upends life as we've known it. We will not be able to reestablish the earth's energy balance that existed for tens of thousands of years prior to the industrial age. In spite of everything we're doing right now and will do in coming decades to lower the rate of global warming (**which will be a lot**), **weather- and climate-related economic and human losses will increase every decade for the rest of this century and beyond.**

Breaking Down the Question

The question: "Will we 'beat' global warming?" really has three distinct or implied parts:

1. Will we drastically reduce CO_2 emissions over a period of three decades? Will we get to zero emissions?
2. How hot will it get? And,
3. Will the increase in global temperatures truly impact the quality of life on planet Earth?

Most of this book deals with the first of these three questions. It's important because the answer has a direct bearing on the relevance of the two follow-on questions. If we (the collective "we") could eliminate global warming emissions quickly and restore the atmosphere to its pre-industrial age composition, the second two questions would become moot. The emissions question also happens to be the one for which my education and experience are most relevant, and therefore, where I may be able to best contribute to the discussion.

The other reason I'm focusing on energy and emissions is that the other two questions are more authoritatively addressed by experts in a variety of different fields. Nevertheless, and with all the appropriate caveats, I will briefly return to these two questions at the end of the book to give you my "answers" to these two issues and the assumptions I made to arrive at those answers.

I am, of course, also making a number of assumptions even in choosing to address the emissions question, including the most basic assumption, i.e., that the emissions question matters.

Revealing My Assumptions

It is a fact that CO_2 gas in the atmosphere interacts with electromagnetic radiation especially in the infrared spectrum, a phenomenon discovered in the late 1800s. This interaction is at the heart of the "greenhouse" effect and the concern about human-induced global warming. Beyond this fundamental agreement, there are a myriad of questions and even "facts" related to CO_2 emissions and their impact on the earth's energy system that some have contested.

How much is the earth warming and to what extent are humans responsible for it? Is there a "saturation" effect for CO_2? In other words, is there a concentration of CO_2 above which adding more CO_2 doesn't cause any additional warming? Has global warming increased the frequency and/or severity of major weather events? Will higher temperatures impact worldwide food production? Will higher temperatures and humidity drive migrations to more human-friendly locations? Will climate changes directly or indirectly increase global conflicts? If changes in the climate do materialize, won't we simply "adapt" to those new climate conditions? What will be the economic impact of global warming and climate change? Over what period of time should we be evaluating changes? How have data gathering methods and technology changed? How can older, less precise measurements of global warming be compared to more sophisticated current measurements? Should we pay more attention to this more recent data? Who should we listen to? Are all the climate researchers working in an echo chamber? How much data do we have to have before we begin to act? How should logic and basic scientific principles inform decisions about what we should do? These are just a few of the many questions that have come to mind as I've tried try to make sense out of the deluge of climate information in technical reports and the popular media.

Climate science is incredibly complex. Our understanding of climate science has expanded greatly over the last 30 years and will continue to expand. Sophisticated new tools for data collection, diagnostic evaluation and modelling have allowed researchers to gather more accurate data and explore fundamental questions more deeply. In addition, the sheer number of researchers working on various aspects of the global warming process and its consequences has exploded. There must be tens, if not hundreds, of thousands of researchers around the world working on untold numbers of topics related to global warming, climate change and its consequences.

The emissions question I'm focusing on in this book is perhaps the easiest of the areas to address but it's important nonetheless because so many people seem to believe we can actually solve the emissions problem by continuing to do what we have been doing. As I will explain, we can't, and until we finally embrace that reality, we won't pursue at least some of the things that could actually lessen the consequences for future generations.

My treatment of global warming and the urgency of the issue **is based on a number of assumptions**, nine of which I list below. The first of these assumptions is critical to the analysis I'm presenting;

the next six relate to the impacts of global warming and are some of the reasons why this battle is worth fighting. The eighth assumption relates to how we "non-experts" consume technical information and the last assumption deals with the timing to take action. If these assumptions (especially #1) turn out to be badly wrong, then the validity and importance of at least some of my conclusions and recommendations could also be called into question. However, in each case, the bulk of evidence supporting these assumptions is strong. My writing accepts these assumptions as fact even if they are not entirely proven in an "academic sense." Here they are:

1. The rate of global warming will increase as more GHGs (greenhouse gases) accumulate in our atmosphere.
2. As warming continues, extreme weather events will become more common. Damaging storms, flooding, heat waves and forest fires will be more frequent or more severe (on average), or both.
3. As warming progresses, food insecurity will increase for a large fraction of the world's population.
4. As global average temperatures increase, large numbers of people may ultimately need to relocate and this will contribute to increased hostilities around the world.
5. The rate of sea level rise will increase as global warming continues.
6. The impact of global warming on people's lives will be significant.
7. A large fraction of the world's population will not be able to adapt to the new climate conditions.
8. Reliance on broadly accepted and collaborated findings from the modern scientific community is not an overly risky proposition.
9. Our knowledge of basic science and global warming processes is sufficient to warrant decisive action.

A few comments on each of the above assumptions are warranted.

1. The rate of global warming will increase as more GHGs accumulate in our atmosphere.

Few if any climate researchers would dispute that the earth is warming and that human activities are the reason for most, if not all, of that warming in the last century. When the natural mechanisms causing changes in global temperatures (positive and negative) are subtracted out, the trend lines show a close correlation between increasing global temperatures and rising CO_2 concentrations. There have been fluctuations over long periods of time, of course, but the National Oceanic and Atmospheric Administration (NOAA) has compiled graphs showing the difference between the measured global average surface temperature and the average of that value from 1900 to 2020. Before 1960, the midpoint for the period in question, that difference was predominantly negative and progressively decreasing in absolute magnitude; after 1960, predominantly positive and growing. This is exactly how the data should display in a warming world. In other words, from 1900 onward, temperatures have been rising due to human activity.

Will the warming continue as atmospheric concentrations of CO_2 rise? I'm assuming it will. There doesn't appear to be a "saturation" effect for CO_2 whereby the earth will stop warming or warm at a slower rate above a certain concentration level. The data showing an increasing rate of warming reinforce confidence in this assumption.

Satellite data in 1996 show decreases in outgoing infrared radiation compared to 1970. This means that radiative cooling from the outer layers of our atmosphere in the range of wavelengths corresponding to GHG absorptions has decreased. GHGs are effectively holding more of certain wavelengths of energy in our atmosphere longer. Perhaps more significantly, surface measurements of downward-directed infrared radiation from multiple studies show an increasing trend (1973 to 2008). In other words, the enhanced exchange of infrared energy between the GHGs in the lower atmosphere and the earth cause the surface of the earth to retain more energy and warm up. The exchange has the equivalent effect of the surface of the earth slipping on a thick down parka. Eventually the heat energy inside leaks out, but in the atmosphere, this process may take years or decades to establish a new equilibrium, and in the meantime the earth inside is overheating -- much as you would eventually overheat inside a super-efficient down jacket.

In trying to make sense out of the totality of global warming reports, my bias is to give greater credence to results relying on more recent data. Data collection instruments, including satellite imagery, have improved the quality and quantity of data being gathered. In addition, greater knowledge about the processes of global warming has improved data collection protocols and identified additional parameters that need to be monitored. Another reason to pay closer attention to more recent data is that it's only been in the last half of the 20th century that the cumulative effects of human activities have really begun to show up. In the sixty years from 1900 through 1960 the atmospheric concentration of CO_2 increased by roughly 35 ppm whereas in the same period of time from 1960 to 2020, the increase was 105 ppm, three times larger.

While it makes sense to give greater weight to more recent data, earlier (much earlier) data can't be ignored. The longer term data spanning thousands, and hundreds of thousands of years, can give us insights regarding natural mechanisms affecting climate cycles and the time frames over which significant changes take place. The changes that give us most concern today are changes that have been taking place over tens of decades, not hundreds of thousands of years. These shorter time frames effectively rule out any long-term cyclical effects from the natural world. Short term events such as massive volcanic eruptions can also be subtracted out of decadal changes. However, the Earth's net energy gain ("energy in" minus "energy out") is a comparatively small difference between two much larger numbers and is, therefore, very susceptible to small changes. As a result, it's reasonable to expect real fluctuations in these trends.

In my analyses and in my writing, I assume that the rate of warming will continue to increase until we stop emitting GHGs and warming will continue until we lower the concentration of GHGs in the atmosphere.

2. **As warming continues, extreme weather events will become more common. Damaging storms, flooding, heat waves and forest fires will become more frequent or more severe (on average), or both.**

Complaining about the weather is a pretty universal human pastime, but anecdotes and recollections aren't "data." There are, however, plenty of academic analyses supporting the observation that "once-in-500-year storms" now seem to be occurring on a semi-regular basis, rainfall amounts are substantially higher in certain areas and flooding events at least in the last decade (notwithstanding increased media coverage and human encroachment into vulnerable areas) seem to be increasing. Certainly the summer of 2022 ultra-high temperatures recorded in England and the catastrophic flooding in Pakistan are well beyond the "norm" of the last two decades. These shorter term trends don't confirm global-warming-driven weather events, but they are consistent with what would be expected. Longer term data show that average nighttime low temperatures and average daytime high temperatures have increased by 2.1° F and 1.9° F respectively over the last 30 years.

Weather is very local. Local conditions can and have given rise to extreme, one-off, temperature events in specific locations at specific times. In other words, a record high temperature in 1913 that hasn't been exceeded in that locations in the intervening 90 years isn't "evidence" that global warming isn't happening. Rather than focusing on temperature extremes, it makes more sense to look at trends that capture more cumulative experiences. Two such indicators are degree-heating-days and degree-cooling-days. The parameter, degree-heating-days, is a measure of how much energy we may need to <u>heat</u> our homes and businesses relative to a fixed baseline temperature. Conversely, degree-cooling-days is a measure of the amount of energy we may need to <u>cool</u> our homes and businesses. From 1900 to around 1960, both average measures were relatively unchanged in the U.S., but after 1960, degree-heating-days have trended significantly down while degree-cooling-days have trended significantly up – trends consistent with measurable global warming.

The characteristics of another type of extreme weather event, a hurricane, may provide further insights regarding global warming. The conditions that give rise to hurricanes are warm, moist air rising from warmed ocean waters. Therefore, it might be logical to expect hurricanes to become stronger as the earth warms. Atmospheric physicists, however, have determined that even under ideal hurricane-generating conditions heat and mass flow limitations, limit the maximum wind speed of hurricanes to around 200 mph. Future hurricanes may not be appreciably more severe than the worst ones we've already experienced, but they may be more frequent which seems to be born out in data spanning over 150 years.

Two parameters, the number of deaths and the amount of property damage are often discussed in connection with major storms, but neither provides particularly useful data in terms of understanding trends in the severity of hurricanes because of other influences. Deaths, for example, have gone down because of better forecasting and communications technology, faster and more effective responses and more robust building standards. Property damage, on the other hand, has gone up because of higher populations and greater economic activity in vulnerable areas.

Rainfall is yet another element of weather. A warming earth also means a warming atmosphere. Because warm air can hold more water vapor, it wouldn't be surprising if precipitation amounts in areas that get rain might be trending upward. Other areas might receive lower than historical amounts of rain due to changing weather patterns. Both effects, higher than normal rainfall amounts in some areas and lower than normal in others are demonstrated in the data. However, aside from precipitation amounts, it may be the changes in precipitation patterns that give us new insights into the effects of global warming and it's these pattern changes to which we may need to pay the most attention.

Given the complexity of weather, the thermal inertia of the land, water and air, the extremely local nature of weather and the sensitivity of "weather" to small fluctuations, trying to tease long-term trends out of weather data is difficult at best and often confusing. Here's what I think matters: **common sense tells me that all the extra energy flowing into our weather-generating heat engine (i.e., our atmosphere and our oceans) as a result of global warming has to go somewhere:** it will show up as higher temperatures, higher humidity, more energetic storms, more frequent storms – or all of the above.

3. As global warming progresses, a larger number of people around the world will experience greater food insecurity.

Since 1960, world food production has been going up even faster than atmospheric CO_2 concentrations. Increased CO_2 does increase the rate of photosynthesis, but crop productivity is a function of much more than just ambient CO_2 concentrations. The suggestion that CO_2 increases have driven these increases in crop yields, however, is not supported by the data. CO_2, temperature, nitrogen in the soil, water, processes unique to different crops, length of the growing season, susceptibility to temperature stresses and many other factors all impact crop yields. From the reports I've read, taken together, the net effect on crop yields of higher CO_2 concentrations and the other changes that come along with it (higher temperatures, more flooding events, effects on pollinators, longer growing seasons, etc.), are likely to range from somewhat negative (initially) to significantly negative as temperatures increase.

The year-over-year yield increases for the major calorie crops (wheat, rice and corn) have been basically flat for several decades. If these yield increases were a strong function of CO_2 concentrations, these increases would form a convex function, i.e., bending upward, mimicking CO_2 concentrations. They don't. In any case, the vast majority of the crop yield increases since 1960 came from the farming practices of the "green revolution" (not to be confused with the "green new deal") including the use of fertilizers, genetically modified crops and improved agricultural practices . . . not increases in CO_2.

I assume that net food production will be negatively impacted by global warming and increasingly so as temperatures rise.

4. **As global average temperatures increase, large numbers of people will need to relocate and this will contribute to increased hostilities around the world.**

Up until recently, economic opportunities, religious freedom, gang violence and war were the primary factors driving migration. As global warming impacts supplies of food and water, survival will become another driver. Will regions of the world less impacted by global warming welcome these climate refugees with open arms? I haven't seen evidence that they will. There have been a few notable exceptions, but even those limited examples of generosity and compassion are likely to fade as the numbers of those seeking "climate asylum" swell in the future. The receiving populations will themselves be struggling with their own global warming consequences. It may be just speculation on my part, but I don't believe climate migration will contribute to global serenity.

5. **The rate of sea level rise will increase as global warming continues.**

Adding more energy to ice at its melting point will increase the rate of melting of that ice and adding energy to water will cause it to expand. The combined effect of ocean water warming and land ice melting will cause sea levels to rise. These facts are not a matter of dispute. The question is: how much of the increasing sea level rise is a result of natural cycles and how much is a consequence of global warming due to excess atmospheric GHGs.

New satellites with sophisticated instruments will soon be available to answer questions about the rate of land ice melting with greater precision than ever before. Separating the effects of natural causes from those attributable to human activities is way outside my areas of expertise. Sea levels were rising <u>before</u> human activities added significant quantities of CO_2 to the atmosphere so clearly there were, and are, natural mechanisms at play. However, the same sea level data show a progressively increasing rate of sea level rise from about 1960 to the present, i.e., the graph of sea level rise after 1960 is convex, bending upward in the same way the atmospheric concentration of CO_2 and global temperatures do. This trend is exactly what would be expected if global warming were a significant contributor to sea level rise.

Land ice melting and the thermal expansion of sea water are most assuredly contributing to sea level rise, but a point that often seems to get lost in these discussions is that we don't need a sea level increase of sixty feet or even six feet before coastal developments and freshwater aquifers are at risk. Increases of six inches combined with a healthy storm surge could (and has) caused massive property damage in low-lying coastal communities in the U.S. and around the world.

Based on data (especially more recent data) and basic science, I assume that as global warming continues, the rate of sea level rise will also accelerate.

6. The impact of global warming on people's lives will be significant.

The impacts of global warming will ripple through the national and international communities. Food and fresh water supplies will come under greater stress, weather events will take a progressively larger toll and require a progressively larger allocation of resources in response, and "climate migration" and global conflicts will increase. In other words, global warming will have both individual and global impacts on people's lives. Many will pay the ultimate price, losing their lives to severe weather events, famine, disease and conflicts exacerbated by global warming. Even those not "directly" impacted will suffer psychological pain knowing what's happening to their fellow human beings.

If we were somehow able to avoid those human tragedies, some have argued that global warming will have little impact on global GDP. That may in fact be true because total GDP is heavily dependent on the sheer number of people contributing to economic activity. However, I suspect that GDP is a less than perfect indicator of the impact of global warming on people's lives.

Consider a boat with a small hole in it. Perhaps only 1% of the people on board will have to be assigned to "bailing duty" to keep the boat afloat. If that hole were larger, an increasing fraction of the passengers would have to begin bailing. Global warming may be like that hole in the boat requiring an increasing outlay of resources to keep the boat from sinking. All those people will be fully engaged performing a vital function and contributing to GDP, but their labors won't boost productivity and contribute to the richness of society. If, as I argue in subsequent pages, more of us will be spending more of our time and resources to "just" survive in a warmer world. As a consequence, humanity as a whole and individual lives will be lesser for it.

7. **A large fraction of the world's population will not be able to adapt to the new climate conditions.**

We are an ingenious species. When faced with an obstacle, we, the collective "we," have always found a way to advance, to cope, to overcome – **and humanity will again this time**. Humanity will survive, but all of humanity will not. How and how much we will need to "adapt" will be as local as the weather. But global warming is a . . . global problem. How will ten or twelve <u>billion</u> of us "adapt?" There are already vast disparities in how resources are allocated across local, national, regional and international boundaries. Those disparities, in a world facing unprecedented challenges, are not likely to shrink. Many of the world's population will simply not have the resources to adapt.

8. **Reliance on broadly accepted and collaborated findings of the modern scientific community is reasonable.**

Given the enormous quantity of data reaffirming global warming and its consequences and the ramifications for all living things, any researcher who could **provide consistent, credible data contradicting basic findings** relating to global warming (or any other fundamental understanding of basic science) would receive an enormous amount of attention. This is an important point in terms of giving those of us, who are not on the front lines of climate research, confidence that mainstream findings can be relied upon.

Think about what would happen, for example, if a paleontologist claimed to have discovered a set of human (or near-human) fossils in a layer of the earth's crust that was home to dinosaur fossils. Such a discovery would, of course, be greeted with more than a healthy level of skepticism, but if proven to be legitimate would quite literally, rewrite human history. In other words, we'd all know about it (and incidentally, the discovery would secure full-tenured professorships for the members of the research team).

Contrarian researchers are rightly exposed to great scrutiny in regard to their methods and findings, but if their work revealed fundamental misunderstandings in "accepted" knowledge, they'd be personally and professionally rewarded. They may take some bumps along the way, but the truth always finds its way home. There will never be a shortage of experts ready, willing and able to challenge common knowledge if they can deliver the goods. This "safeguard" against "group think" should give us all some comfort that accepting the common consensus that emerges from modern, peer-reviewed science is usually not an overly risky proposition.

9. Our knowledge of basic science and global warming processes is sufficient to warrant decisive action.

Our knowledge of global warming and its consequences have expanded exponentially in the last two decades. It may not be fully complete, but the prudent course of action is to use the information we do have, extrapolate into the future and act accordingly. This is especially important because the potential consequences are so great, the problem is increasing every day, a credible response will take so long to implement and the time available to act is slipping away. We may not know with certainty how rapidly land ice is melting, but we know that an increasingly warm environment will accelerate that process. We may not know how many more energetic storm events will be spawned by global warming, but we can say with a very high level of confidence that adding more energy to our weather-generating system won't make our weather "more benign." We may not know exactly how food crops will be affected by a warmer, higher CO_2 environment, but we know that even infrequent extreme temperature events can take an enormous toll on plants (and people) and that holding onto the yield improvements achieved in the last 50 years is going to be critical to feeding the world's growing population. We don't have to dot all the "i's " and cross all the "t's" before we act.

I'm reminded of that old adage, "If it looks like a duck, quacks like a duck, walks like a duck," (and in this case, you're inside an enclosure at a duck farm with a sign that says, "Beware of the Ducks," that animal you're looking at. . . it's probably a duck).

Nature - The Perfect "Integrator"

Scientist and environmentalist James Lovelock developed a concept he called the Gaia Hypothesis which describes our earth as a giant, self-regulating "organism." I think of our earth in a similar, though less poetic way, as a gigantic array of interconnected feedback loops with the whole array behaving in accordance with invariant physical laws. ["A gigantic array of interconnected feedback loops" – that's an engineer's way of saying: "If a butterfly flaps its wings in the Amazonian rain forest, it can change the weather half a world away."]

Nature is the perfect "integrator." It takes all its inputs big and small, processes them, and churns out a response. Sometimes Nature's

> *Our earth functions like a gigantic array of interconnected feedback loops. Those interconnections make Nature the perfect "integrator." It takes all of its inputs, big and small, processes them, and churns out a response. Nature's results are always real and always true.*

response is to our liking, sometimes not, but Nature (the physical world) doesn't care how we feel. It just is. Denying a physical reality may be expedient for some, but that doesn't make the denial true. Nature's results are always real and always true.

Politicians and even some climate action activists seem to be characterizing the fight against global warming as a sprint to the finish line. It's not. It'll be more like a marathon -- a never-ending, uphill marathon.

Authoritative national and international governmental agencies are also fond of telling us that the window for meaningful action to combat global warming "is closing." They're wrong, but depending upon what they are trying to communicate, they're wrong in two very different ways.

If they mean that after a certain time or after a specific elevated concentration of CO_2 in the atmosphere is reached, efforts to reduce CO_2 emissions will be useless, they're wrong because everything we do to reduce CO_2 emissions from this day forward will slow the rate of global warming compared to what it would have been had we not made the effort. This is true even if we never get to zero carbon emissions, much less return the atmospheric concentration of CO_2 to pre-industrial age values.

If, on the other hand, they are saying that we still have a chance to reduce CO_2 emissions to levels that will allow us to avoid severe global warming consequences, they're also wrong.

Factoring in the dynamics of what it would have taken to make that change, that **window is not *closing* -- it effectively clos_ed_** (past tense) **decades ago.**

We are, in effect, conducting the largest-ever global experiment. It's as if we said, "I wonder what would happen if we doubled the concentration of CO_2 in the earth's atmosphere over a period of 150 years? Unfortunately, we're going to find out.

The Layout of the Book

This is not a conventional book:

1. Rather than coming at the end, a chapter summarizing my conclusions appears after the introductory chapters and **before** any of the supporting material. I've organized it this way so that **you, the reader, will know exactly where I'm going with my arguments and can be on the lookout for any logical inconsistencies.**

2. I've intentionally written in a conversational style with only a modest amount of quantitative data to make the material as accessible as possible.

3. In some cases, I've used sentence fragments, bolding, punctuation, and repetition to make a point. Sometimes those aspects of my writing are intentional. Please don't let the many imperfections you may find get in the way of my message. (With your help I'll fix the errors in the second edition).

4. I don't claim to be an expert in climate science, weather, climate modelling, biology, oceanography, or anything similar. (That's why I detailed earlier in this chapter the assumptions I'm making). The climate disciplines were pivotal when the question was: "Is global warming real?" The answer to that question is no longer in doubt. The critical question now is: "What can we do about it?" To address this question, my experience in energy R&D, the electric utility industry, electric generation planning, and an MBA, give me what I hope is a unique platform from which to contribute to the global warming discussion. (A brief summary of my education, experience, values, **and biases** is provided in Appendix B).

5. The book isn't finished. The optimist in me envisions a second edition which will include technology updates along with input, comments, corrections etc. that are solicited from you, the readers of this edition.

At least this first edition is free -- well, almost free. I'm pledging all net royalties plus a matching amount up to $10,000, annually, to humanitarian and environmental causes. (Hey, if Bill Gates can give away the royalties from his global warming book, so can I.)

This book focuses on the catastrophic global imbalance (CGI) issue mostly from the perspective of the United States, and to a lesser extent, other industrialized countries. This is because I'm naturally more familiar with the data and issues as they relate to the U.S. and our energy grid. Another reason to take this narrower perspective is that rich Western nations, most especially the U.S., are contributing the most to the problem on a per capita basis. As many other authors have pointed out, it will be the comparatively less developed countries and their inhabitants who will be most impacted by the coming changes in the world's climate, but it's the industrialized countries where the biggest changes in the production and use of energy need to be made. It's also these developed countries that have the resources to make those changes and where technologies to address the problem will continue to be developed and refined.

In addition to focusing mainly on the U.S., this book also focuses on us humans. We're the ones who are causing global warming and we're the ones who will have to transform our energy economy, but global warming and climate change obviously impact all living things, plant and all animals, not just humans.

I hope you'll read the chapters of this book in order, but if you want to go straight to the bottom line, start with the Fifteen "Myths" in Chapter 21, the "Quotable Quotes" in Appendix A, and the closing argument in Chapter 22. If you read these sections first, I hope you'll then circle back to review the supporting material.

At the end of certain chapters there's a section called "Action Items." These "Action Items" are recommendations related to the topic in that chapter. These are some of the things we need to do to slow down our relentless march toward climate disaster. Everybody who writes about global warming urges their readers to get involved, to write letters or make calls to their elected representatives. **These "Action Items" are the things that I'm suggesting you should demand that corporate, religious and especially governmental leaders do to make a credible assault on global warming.**

As you read the book, I want you to repeatedly ask yourself: "Does this make sense to me? Do I agree or disagree with this point?" Start by make a mental note of your answer to the central question the book poses: "Do you think we'll 'beat' global warming?"

I envision a range of possible answers:

1. Yes, I hear about big renewable energy projects nearly every day and all the major auto manufacturers are going electric.
2. Yes, solar and wind power are not only cleaner, they're also cheaper. We'd be cutting back on fossil fuels even if global warming weren't a problem.
3. Yes, I know we're not doing enough today, but the U.S. has rejoined the Paris Climate Accords. Climate negotiators just have to hammer out an agreement.
4. Yes, I'm certain the solutions to global warming are in the R&D labs right now.
5. Yes, when there's a buck to be made, some very smart people will get very busy, very quickly.
6. No, CO_2 levels are going up every year. I know it's not good, but I'm not sure what we can do about it.

Global warming is a multi-dimensional problem. Any meaningful response to the challenge must involve not only technologies, but all the other factors that affect how we'll respond. I touch on each of them in the chapters that follow.

Chapter 3 provides a brief overview of the science of global warming. Chapter 4 describes energy production and usage, what it would take to adequately address the problem and why we're not going to do what needs to be done. This is the "conclusions chapter" that comes before the supporting discussion.

Chapter 5 makes the case that how we respond to this crisis is important even if we're not going to "beat" it.

After these introductory chapters, I present the supporting arguments, broken down by energy sector in Chapters 6 through 15. Chapter 16 discusses some of the technological fixes that are often cited as the silver bullets to "beat" global warming. (Sorry, there are no silver bullets).

In Chapter 17 I narrowly address the role of the media. Chapter 18 looks at what we would need to do to slow the relentless increase in GHGs. Chapters 19 and 20 imagine how the unfold. (It's not pretty).

In Chapter 21, I list 15 common "myths" relat warming and the solutions we're pursuing. In Chapter 22 I present my closing argument. The appendices provide a list of quotes from the book and a brief bio of yours truly.

Throughout this book, you'll find the notation: [*Input]. This notation identifies issues or topics on which I'm specifically soliciting additional material from readers for inclusion in the second edition. Any expert contributions, dissenting opinions or other comments intended for inclusion in the next edition may be channeled through the website: www.BlueOasisNoMore.com. In the second edition, I will list myself as "author/editor" (rather than author alone) in order to highlight what I hope will be a wide range of credited input provided by others. Readers should not feel limited to providing input or comments on just those areas where I've specifically requested [*Input]. I invite all constructive input addressing the question of whether or not we can "beat" global warming.

What's In a Name?

Just a word about the title of this book. Yes, I'm aware that the color of our planet might be better described as "blue-green," but the title "Blue-Green Oasis No More" just didn't sound right. For me at least, it would have focused you, the reader, on an insignificant detail, i.e., a color, and not on the most important words in the title, namely, "No More."

I considered some alternative titles. One, "The Canary is Coughing," was a reference to the 20th century practice of using caged canaries in coal mines as carbon monoxide detectors. When the canary stopped singing or fell over dead, the miners would know that odorless, deadly gases had built up in the mine and they needed to get out fast. Today, a colorless, odorless gas is building up to concentrations in our atmosphere that will have deadly consequences. The evidence is all around us and scientists are providing stark and urgent warnings, but we're acting as if the canary is still singing. To describe the canary as "coughing," might imply that there's still time to "beat" this thing. There isn't.

Another early working title was "Unavoidable Catastrophe." The problem with this possible title is that when people normally think about a catastrophe, they often think about **a catastrophic event**, a point in time. That's not what climate change is. Global warming and climate change will be persistent and slow when measured against the life span of a human, and "lightning fast" when measured on a geological time frame.

Another possible title was "Drill Baby Drill: From an Inconvenient Truth to an Unpleasant Outcome." This seemed just too political for the main message of the book which is inherently apolitical. I am hoping this book can reach readers of all political persuasions independent of my clear biases, which I haven't tried to obscure.

The Sting of Global Warming

Governor Inslee statement quoted at the beginning of this chapter, was only half right. He was correct in pointing out that we may be the first generation to feel the "sting" of climate change. The scientific data along with anecdotal evidence in the form of more severe and more frequent weather-related disasters have been accumulating from Maine to California and from Inslee's home state of Washington to the Florida Keys, and everywhere in between.

I may be putting too fine a point on this, but for me, the second half of Governor Inslee's statement is problematic. He is completely correct in trying to convey a sense of urgency, but technically, ours is **not** the last generation that can do something about global warming. In fact, **our generation, and every generation from this time forward, will be coping with the consequences of global warming** – and will be trying to "do something about it." As I hope this book makes clear, this will not be a "one-and-done" rescue. It took us decades to get into this mess, and we'll be fighting its consequences, not for decades, but for centuries.

The other way in which the second half of Governor Inslee's statement is problematic is that the last generation that might truly have had a chance of doing something about climate change would have been his parents' generation starting in the 1950s or '60s. At

> *The options that are realistically available to us and the time available to us to "do something about it" are not sufficient to the challenge.*

that time, however, only a handful of people had even uttered the words "global warming," and with the exception of nuclear power, most of the technologies that we will be using to combat global warming today either didn't yet exist or were in their infancy. The technologies that are available to us even now, however, and the time for us to act to "do something about global warming" are not sufficient to the challenge.

More Than Technology

While technology is the element of the problem that most people focus on, our ability to adequately respond to our catastrophic global imbalance is not exclusively a function of technology. **It's not even primarily a function of technology.** Economics, politics, education, psychology, culture, emotion, international relations . . . and technology all have a bearing on how we will respond to global warming.

In the course of this book, I hope to make clear:

We're **DUMPING** unfathomably large quantities
of CO_2 into the atmosphere.

We **NEED** to reduce those emissions to **ZERO** very **QUICKLY**.

If we were able to somehow stop making the problem worse by eliminating all human-caused CO_2 emissions,

THE EARTH WILL CONTINUE TO HEAT UP.

WE ARE DOING A LOT to reduce those emissions
and
WE WILL DO A LOT MORE in coming decades.

BUT, we can only

DIMINISH OR DELAY

the worst impacts of global warming.

WE WILL

NOT DO ENOUGH, QUICKLY ENOUGH

TO "BEAT"

GLOBAL WARMING. Full stop.

3 HOW WE GOT HERE

"It's not nice to fool [with] Mother Nature."

**Variation on the TV Commercial
For Chiffon Margarine, 1974**

Before discussing the things we would need to do to "beat" global warming, it may be useful to briefly review why global warming is happening. Doing so requires turning the clock back . . . **way back**.

In The Beginning . . .

Asteroids, volcanoes, internal heating, and many other factors made our earth an especially dynamic and inhospitable place for its first couple of billion years. Even after asteroids stopped pummeling the earth on a regular basis, our planet continued to be a tumultuous place. Eventually primitive plants began using the energy of the sun to power the process of photosynthesis. Plants extracted CO_2 from the air, discharged oxygen as a waste product and altered the composition of our atmosphere in a way that made all animal life possible. Oxygen-breathing animals from the single-cell variety to woolly mammoths completed the carbon cycle. They used oxygen in metabolic processes to convert plant and animal material into energy (i.e., "burning" calories) with CO_2 as a by-product.

Conditions on Earth in those early years were ever-changing and not always in ways that were favorable to life. Those changes included slow, but dramatic changes in the chemical composition of the atmosphere. Scientists estimate that a few of those changes may have snuffed out as much as 80% of all life on earth when they occurred. In addition to the changing chemical composition of the atmosphere, the earth's land masses were literally drifting around the globe.

Fast forward hundreds of millions of years to a time when life is flourishing in every corner of the earth. As plants and animals lived

27

and died, their remains, including compounds containing hydrogen and carbon, accumulated and were covered over. In time, some of those remains were transformed by temperature and pressure into deposits of coal, oil and natural gas, i.e., "fossil fuels." (I told you this was going to be brief and superficial!)

There for the Taking

Deposits of carbon and hydrogen compounds remained sequestered underground in geologic structures for eons until an amazingly ingenious animal species came along. That extraordinarily successful species was us.

The earliest humans figured out how to use fire. At first, they burned the hydrocarbons all around them (trees and brush) for heating, cooking, and light. Modern man, however, took hydrocarbon consumption to a whole new level after discovering energy-dense deposits of coal. They were there for the taking, the generous bounty of Mother Earth. We began tapping into these reservoirs of energy as the industrial age took hold in the late 1800s. Over time, our exploitation of fossil fuels expanded from coal to oil and then natural gas – and along the way, became a dependence, an addiction.

There are many grades and compositions of coal, oil, and natural gas. The energy-beneficial part of each fossil fuel is locked up in the compounds of carbon and hydrogen. For the sake of simplicity, I'll highlight just the simplest and cleanest of them, natural gas, but the basic story is the same for each.

The chemical composition of natural gas is CH_4. The formula for the combustion of natural gas is:

$$CH_4 + 2\,O_2 \;\longrightarrow\; CO_2 + 2\,H_2O + \textbf{Energy}$$

From the human perspective, the desirable part of this chemical reaction is the **energy** released in the process. That energy is why we burn fossil fuels. However, in addition to the energy released in the process, one of the byproducts of combustion is carbon dioxide, CO_2.

Carbon "Resurfaces"

Prior to the industrial age (i.e., for most of the time our species has been walking the face of the earth) mankind had virtually no impact on the chemical composition of the atmosphere. There were two reasons for this. First, there weren't many of us around. Second, our pre-industrial age ancestors were burning plant life, logs and brush which were readily available all around them. The carbon in those fuels was being recycled from the atmosphere to plant and animal life and back into the atmosphere on a short cycle length.

The fossil fuels we're burning today are different. The CO_2 we're releasing by burning fossil fuels has not been in the atmosphere for eons. It's not new to the earth, but it's "new" to our present-day atmosphere.

Once released, most of that CO_2 remains in the atmosphere for a long time. Researchers typically talk about an atmospheric half-life for CO_2 of between 50 and 200 years. Researchers at NASA's Jet Propulsion Laboratory have characterized CO_2 as "hang[ing] around in the atmosphere for between 300 and 1,000 years." In other words, nearly all of the CO_2 being released into the atmosphere today will still be causing global warming and ocean acidification, not for years and not for decades, but for centuries. This is a critical point – the effects of the CO_2 we're

> *The buildup of CO_2 in the atmosphere will continue to impact the earth's heat balance not for years, and not for decades, but for centuries.*

releasing into the atmosphere today will be felt for **centuries.**

The concentration of CO_2 in the atmosphere is measured in ppm, parts per million. Prior to the industrial age, the concentration of CO_2 in the atmosphere had been fairly steady in the range of 260 to 280 ppm. For hundreds of thousands of years before the industrial age, that level had fluctuated between roughly 200 and 280 ppm and changed only very gradually.

One observable, predictable, and natural fluctuation in atmospheric CO_2 concentration is seasonal. The concentration of CO_2 in the northern hemisphere typically goes down in the spring and summer months, (i.e., the growing season), when plants are extracting CO_2 from the air and growing. The atmospheric CO_2 concentration goes up again in the fall and winter as the plant products are consumed or die and decay. This seasonal fluctuation in the atmospheric CO_2 concentration is between plus and minus three ppm.

For the last several thousand years, the **non-seasonal** concentration of CO_2 in the atmosphere fluctuated very slowly in a fairly narrow band.   Or, at least it **used to** . . .

And Then . . .

As industrial activities ramped up, ever greater quantities of CO_2 accumulated in the atmosphere from the burning of fossil fuels. As scientific knowledge grew, people began living longer and (at least in some ways) healthier lives. Populations increased. Larger numbers of people gave rise to more industrial activity and with it the burning of even more fossil fuels.

As noted, most of the CO_2 released into the atmosphere remains there for a very long time. Between a third and a half of the CO_2, however, is ultimately absorbed by the oceans. While this process decreases the rate at which CO_2 builds up in the atmosphere, it causes another problem: ocean acidification. This process decreases the ocean concentration of carbonate ions, a chemical building block for ocean species that create shells. Ocean acidification along with ocean warming have been measured and their impact, such as coral reef die-offs, have been observed with greater frequency in recent years.

Most of you have no doubt seen the graphs of the atmospheric concentrations of CO_2 that were prominently featured in the documentary film *An Inconvenient Truth*. Similar curves are shown in the next two exhibits. Exhibits 3.1 shows the recent upward trend and seasonal changes in CO_2 concentrations. Exhibit 3.2 shows the longer-term concentrations.

Exhibit 3.2 shows the dramatic increase in atmospheric CO_2 concentration driven by the skyrocketing use of fossil fuels in the industrial age. No other causes that could affect CO_2 concentrations by this much were in play in the period from the late 1800s to the present. [These data are readily available on the Internet. Google: "CO_2 concentration graphs" and take your pick. Pat Linse redrew these specific graphics and are used here by permission of the artist and in her honor. They were part of a twenty-eight-page paper entitled *A Skeptic's Guide to Global Climate Change* by Donald Prothero, distributed by the Skeptics Society. It's a concise and informative document that provides excellent background material on the global warming issue and specifically addresses counterclaims by climate change deniers.

Exhibit 3.1 Recent Atmospheric CO₂ Concentration

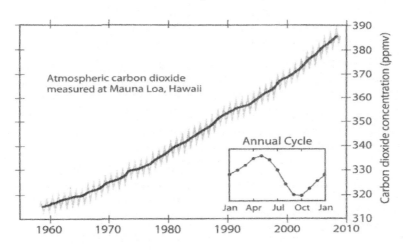

Exhibit 3.2 Long-term Atmospheric CO₂ Concentration

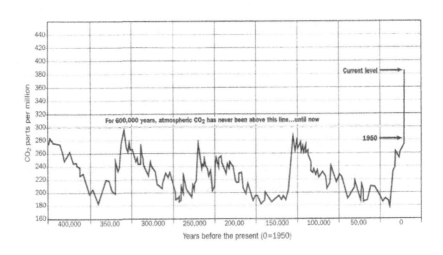

The Only Constant is Change

We've all heard the expression "the only constant is change." We also know that change is often disruptive. The two important quantitative characteristics of any change are the **rate of the change** and **the magnitude of the change**. Taken together, these two parameters determine how disruptive a change may be. If the magnitude of a change is small, the change will probably not be disruptive even if it's fast. If the magnitude of a change is large but takes place

> *The two most important things about change are: the rate-of-change and the magnitude of the change.*

over a long period time (e.g., hundreds of thousands of years), the change is also less likely to be disruptive. The most problematic changes are the ones that are large and fast.

The change in the atmospheric concentration of CO_2 that's taken place in the past 120 years (Exhibits 3.1 and 3.2) has been both large and fast. That momentous change has given rise to global average temperature increases that are not overly large in absolute terms, but easily large enough to wreak havoc in the world of living things and on the earth's energy pathways (ocean currents, the water cycle, and air currents). Rachel Carson made this point in her book, *The Silent Spring*:

> "It took hundreds of millions of years to produce the life that now inhabits the earth -- eons of time in which that developing and evolving and diversifying life reached a state of adjustment and balance with its surroundings. Given time -- time not in years but in millennia -- life adjusts, and a balance [is] reached. For time is the essential ingredient; but in the modern world there is no time."

We can't know all the impacts of global warming, but the ones we do know are beyond consequential. They include: increasing average global temperatures, extreme heat events, extreme humidity conditions, disrupted weather patterns, more frequent storms, more severe storms, more flash floods, higher coastal storm surges, crop failures, forest fires, grassland wildfires, habitat destruction, ocean acidification, marine food chain disruptions, animal extinctions, melting ice masses, sea level rise, etc. All of these deteriorating conditions in the natural world have grievous consequences for the family of man. And again, what makes those changes so devastating to all living things are the **rate of change** and the **magnitude of those changes**.

Compared to other animals, humans are incredibly adaptive and ingenious. We can usually adapt to large and rapid changes, but the damaging effects of global warming will be felt by many other species long before they reach a level that we can't tolerate or to which we can't adapt. That's exactly what's happening. Nature is trying to get our attention. We just need to listen.

> *"Nature" is trying to get our attention; we just need to listen.*

Besides our ability to adapt, another reason we haven't been heeding nature's SOS it that global warming and climate change are consequences of a large and multi-faceted dynamic system with built-in time delays. These delays make it hard for us to connect "causes" with "effects." We are accustomed to viewing the world in static "snapshots" when in reality what's unfolding around us is an interactive "movie."

A Fragile Balance

In the modern era, the atmosphere has become our gaseous garbage dump. That's especially unfortunate because besides being the source of the air we breathe, this thin layer of gases performs a variety of other functions critical to sustaining life on earth. Ninety-nine percent of the earth's atmosphere is concentrated in a thin shell just twenty miles thick. This gas layer is just 0.0025 times the diameter of the earth. This extremely thin layer of gases and the earth's magnetic field, shield the earth from a significant amount of electromagnetic and other radiation from our sun. Our atmosphere also largely determines the heat balance of our planet.

The sun's electromagnetic energy arriving at the outer edge of the earth's atmosphere consists of a range of different wavelengths (i.e., visible light, ultraviolet light, infrared waves) Each of these wavelengths interacts differently with the different chemical constituents of our atmosphere and earth. The land and oceans absorb roughly half of the energy impinging on the earth; the different components in the atmosphere absorb a little under a quarter. The remaining quarter is reflected directly back into space. Ultimately, nearly all of the energy absorbed by the land and water is re-radiated in the infrared spectrum somewhere in our atmosphere.

When the amount of energy absorbed inside our atmosphere is equal to the amount of energy escaping back into space, the temperature of the earth is stable. When they're unequal, the planet either heats up or cools down, which it has done many times in its long history.

Our atmosphere acts as a transparent medium to most of the sun's incoming energy. The incoming waves that aren't reflected back into space or absorbed, pass through the atmosphere and are absorbed by the land and water. When those electromagnetic waves are absorbed, their energy is re-radiated in the infrared spectrum. The two primary components of our atmosphere, nitrogen, and oxygen, don't absorb infrared waves. If our atmosphere consisted of just these two gases, all of the re-radiated infrared energy would escape out into space. But other components in our atmosphere, including water vapor, CO_2, and other gases can absorb the energy of the re-radiated infrared waves, trapping it at least temporarily inside our atmosphere. This is known as the "greenhouse effect."

In an actual greenhouse, the transparent medium is the glass that makes up the walls and roof of the structure. Incoming sunlight passes through the glass, is absorbed by the plants and soil inside and the energy is ultimately re-radiated in the infrared spectrum. The energy in infrared waves can't pass back out through the glass; the glass blocks it. The energy is trapped inside, and the greenhouse warms up which is what greenhouse owner wants to happen.

In our atmosphere, water vapor, CO_2 and other gases act like the glass in a greenhouse. That's why they're known as "greenhouse gases," GHGs.

CO₂ Has Accomplices

In addition to water vapor and CO_2, there are a number of GHGs that allow solar radiation to be absorbed by the earth and prevent re-radiated infrared energy from escaping out into space. As it turns out, CO_2 is one of the **least** effective players in regard to trapping energy. Relative to CO_2, many other gases have a greater ability to trap energy in our atmosphere. This heat trapping effectiveness is known as Global Warming Potential, GWP. GWP is a function of how long the gas persists in the atmosphere and how it interacts with different wavelengths of radiation. From the perspective of limiting global warming, suffice it to say that a lower GWP is better than a higher GWP.

Exhibit 3.3 shows some of the global warming gases and their relative effectiveness per pound in terms of trapping energy in our atmosphere.

Exhibit 3.3 GHGs, Persistence, Relative GWP and Impact %

Greenhouse Gas (GHG)	Lifetime (Years)	100-Year GWP per pound	% Contribution To Global Warming
Carbon Dioxide (CO_2)	Hundreds	1	75
Methane (CH_4)	12.4	25 - 28	17
Nitrous Oxide (N_2O)	121	298	5
Hydrofluorocarbon-23 (CHF_3)	222	14,800	Small
Sulfur hexafluoride (SF_6)	3,200	22,800	Small

One of the important "takeaways" from Exhibit 3.3 is that efforts to counteract global warming need to focus on all major GHG emissions, not just CO_2.

> *Efforts to counteract global warming need to focus on all sources of GHG emissions, not just CO_2 from power plants and automobiles.*

The most important GHG, H_2O vapor, is omitted from this table because we have little or no control over its presence in our atmosphere. Water vapor is naturally occurring in the atmosphere and has both a positive and a negative effect on the earth's heat balance. As high white clouds, water vapor reflects more incoming solar radiation back into space but it also has the opposite effect of preventing infrared radiation emitted close to the surface of the earth from escaping.

Methane has a much higher GWP than CO_2, as do all the other gases listed in Exhibit 3.3. However, in spite of its higher GWP, methane is responsible for a much smaller percentage (17%) of global warming. This is because the quantities of methane emitted due to human activities are so much smaller than CO_2 and because methane has a much shorter residence time in the atmosphere. This doesn't mean we can ignore methane. Efforts to reduce methane releases from natural gas extraction and processing, livestock operations, landfills and manure management are vitally important in combatting global warming.

There's an additional worrisome methane issue looming on the horizon. As the earth warms, the permafrost below the topsoil in the Arctic tundra is thawing. As it thaws microbes attack organic matter and release methane and CO_2. The continued warming of the planet will cause expanses of tundra to thaw faster, releasing more methane into the atmosphere which will cause even more warming. This is a classic "positive feedback loop" that's anything but positive for the planet. The PBS NOVA episode, "Arctic Sinkholes" describes the extent of the threat and provides stark evidence that it's already happening.

The reason why most authors (including myself) focus on CO_2 and the consumption of fossil fuels is that the quantities are so large and excess CO_2 in the atmosphere is caused predominantly by human activities. Even with its lower GWP, CO_2 is responsible for 75% of the added heat absorption by the earth.

The components that make up the mass of the earth (water, land, ice, atmosphere, living things, etc.) all have different heat capacities, different densities, and different abilities to exchange energy with one another. The combination of these physical characteristics contributes to a lag time between the addition of energy to a multi-component body and the temperature rise that body exhibits. As the earth absorbs excess energy, the temperature of the earth rises. But just as water heated on a stove takes time to boil, the heating and cooling of the earth also take time. The mass is said to have thermal inertia. This is why the hottest days of summer don't typically coincide with the longest day of the year when the number of daylight hours is the greatest. The hottest days typically occur a month or two later. There's a "lag" in the system. Similarly, the coldest days of winter typically occur a month or two after the shortest day of the year. The oceans and other large bodies of water with their high density and high heat capacities have high thermal inertia. They take a long time to reach thermal equilibrium with their surroundings and are especially effective in moderating local temperature swings. They act as a buffer.

In addition to water and land, ice is another component of the earth's thermal buffering mass. Not only does ice have a much higher heat capacity than air, ice also goes through a "phase change" from a solid to a liquid when it melts. This phase change requires the absorption of a large quantity of energy. It takes the same amount of energy to melt a pound of ice (at 32° F) as is required to heat that same pound of liquid water from 32° F to 176° F. (As a point of reference, the water coming out of your water heater is typically around 130° F). In other words, **it takes a lot of energy to melt ice.** This is one of the reasons why glaciers and ice sheets are especially slow to melt. This is also why it takes a surprisingly long time to melt the ice that can build up on the cooling coils of a refrigerator, even using a hair dryer on the hottest setting.

The temperature moderating effect of the earth's surface ice melts is being lost. Melting ice is sucking energy out of the ambient air that would otherwise show up as higher average air temperatures. This means that some of the excess energy that's being absorbed by the earth as a result of global warming isn't yet showing up as higher global temperatures. Over the next few generations, the moderating effect of the earth's ice masses will have melted away, literally, giving rise to more dramatic temperature swings.

And This Heat Balance Is Important Because . . .

Prior to the industrial age, the energy flow into and out of the earth was in balance, but as the consumption of fossil fuels ramped up, the concentration of CO_2 in the atmosphere began to climb. Even though CO_2 constitutes only a small fraction of the gases in our atmosphere, it has an outsized impact on the earth's total heat balance.

Over the last 150 years, human activities (primarily burning fossil fuels) have caused the atmospheric concentration of CO_2 to increase from 280 ppm (part per million) (0.0280 %) up to 420 ppm (0.0420%), **a 50% increase!** Exhibit 3.4 showed that increased concentration of CO_2 in the atmosphere.

The warming of the planet over the last 150 years (Exhibit 3.4) correlates closely with the dramatic increase in the atmospheric concentration of CO_2 shown in Exhibit 3.2. Because we know CO_2 and the other GHGs absorb infrared radiation more effectively and prevent more of it from escaping the atmosphere, the connection between Exhibit 3.2 and 3.4 is more than a correlation. With an understanding of the physics involved, it demonstrates a causation.

Exhibit 3.4   1000 Years of Temperature Data

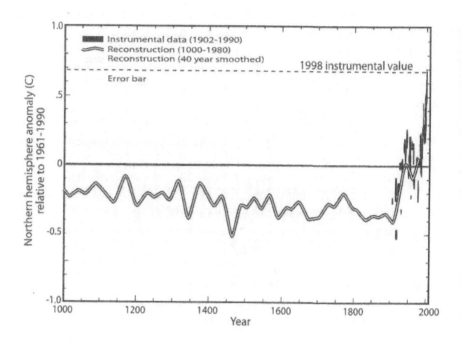

Just to Recap

Human activities, specifically the burning of fossil fuels, have driven up the atmospheric concentration of CO_2. This rapid change in the composition of the atmosphere is causing the earth to absorb more energy and the oceans to become more acidic. The global average temperature is increasing with what are likely to be devastating consequences for all living things.

We're "SLOW ROASTING" our planet.

So, Now We're All on the Same Page, Right?

Well, not exactly. A small (but fortunately decreasing) percentage of people are still not persuaded by the data, anecdotal evidence and what they can see and feel with their own senses. From my experience it seems as though those who deny that global warming is real can be grouped into several categories:

1. Those who truly don't understand the science or the data. They are unwilling or unable to invest the time to understand it and choose to reject a concept they don't understand. Vested interests in the fossil-fuel industry are doing what they can to keep putting doubt in the minds of individuals who reject global warming. For those vested interests, it's all part of a high-stakes game for power, influence, and money, and they're more than willing to pray on the emotions, lack of knowledge or contrarian orientation of their target demographic.

2. Those who agree that global warming is real but reject that it's caused by humans. They oppose mitigation efforts because they don't believe we can control something we didn't cause.

3. Those who know that it's real and caused by humans but claim otherwise because they oppose doing anything about it. Recognizing how expensive and disruptive it would be to meaningfully combat global warming, they're unwilling to make sacrifices in the "here and now" for a future benefit when they won't be around. For them, denying that global warming is real gives them some "cover." By claiming that global warming isn't real, they think they can avoid being seen as so selfish that they don't care about future generations. This is probably a big enough group to warrant its own acronym: they're the NIMTOEs, the "Not In My Time On Earth" group. They don't care what happens to the planet because they're not going to be here.

4. Closely related to the NIMTOEs are those who deny that global warming is real because they are trying to protect some financial enterprise that would be adversely affected by policies to curtail CO_2 emissions. A politician from a coal-producing state or a coastal land developer, are two examples. They claim it's not real because acknowledging that it is, would hurt them politically or financially.

5. Those who acknowledge that global warming is real but are convinced that a soon-to-be-discovered technology or an omnipotent God will rescue us from disaster. This group opposes mitigation measures because they're certain a "silver bullet" is being developed in a laboratory somewhere or that the situation will be remedied by a higher power.

6. And finally, there are those some accept that global warming is real and caused by human activities but claim that the consequences of global warming and climate change won't be catastrophic. In spite

of evidence to the contrary that they can see with their own eyes, they use this reasoning as the rationale for refusing to support countermeasures.

Some deniers claim that the anomalies in weather patterns and record high temperatures are merely a statistical fluke. They are correct that random events can sometimes produce unexpected outcomes. It is statistically possible, for example, to roll "snake-eyes" eight times out of ten rolls of the dice, but if you're sitting at a table where that's happened, you should get up and walk away . . . as quickly as you can.

Statistically, it's possible that eight of the ten hottest years of the last century occurring in the last decade could be a statistical fluke but believing that would require intentionally disregarding accepted science related to the interaction of electromagnetic radiation (sunlight) and matter (our atmosphere). The recorded high temperature data are not the result of random events; they are the result of physical laws.

Whatever the rationale or flawed reasoning, I would pose a series of questions to deniers in an effort to identify the true root of their skepticism:

1. Do you accept that CO_2 is a by-product of burning fossil fuels?
2. If so, where is all the CO_2 that's being produced by burning fossil fuels going if the majority of it isn't ending up in the atmosphere and oceans?
3. If you reject the idea that human activity is responsible for the increase in atmospheric CO_2, then what's your explanation for where the added CO_2 that's being measured in our atmosphere is coming from?
4. Do you understand and accept the science that increased CO_2 concentrations in our atmosphere have a greenhouse effect, i.e., that these gases trap additional energy from the sun?
5. If you accept the validity of the greenhouse effect of CO_2, where is the additional energy that's being absorbed in the atmosphere going if it's not the cause of the measured increases in global average temperatures?
6. Do you accept the fact that water and air will increase in temperature when they absorb energy and that they will heat up faster as the rate of energy absorption increases?

7. Do you accept that land ice will melt at a faster rate as global average temperatures increase?

8. Do you accept that water from melting ice on land can and will flow into the oceans causing ocean levels to rise?

9. Do you know that raising the temperature of water causes the water to increase in volume which in turn contributes to rising sea levels?

10. Do you understand that increasing that all the extra energy being absorbed in our atmosphere and oceans has to go somewhere and that more severe storms and altered weather patterns are a likely consequence of that added energy?

11. If you understand and accept all these things as true, how can you deny that global warming is real, is caused by human activity and will have significant consequences for mankind?

The burden of proof is no longer with those who are raising the alarm about global warming. The deniers, those who still have doubts, or claim to have doubts, need to provide proof to the contrary.

The Path Forward

An ever-increasing majority of people accept that global warming is real and that we need to do something about it. Since we don't have a dimmer switch for the sun (geoengineering proposals will be discussed in Chapter 16, Technological Fixes), and because the technologies to extract CO_2 from the atmosphere (carbon capture and storage, also discussed in Chapter 16) are currently expensive and would be completely ineffective if we keep dumping CO_2 into the atmosphere, the only viable near-term strategy is to "stop the bleeding," i.e., stop dumping excessive quantities of CO_2 into the atmosphere.

That's what most of this book is all about – **transitioning away from a worldwide fossil-fuel economy to a worldwide carbon-free energy engine.**

> *Combating climate change will require much more than solar panels, wind turbines and EVs.*

Most people think of this as a "renewables" economy, but it's so much more than just solar panels, wind turbines and electric vehicles (EVs).

In the next chapter I'll review what we need to do to stop global warming. I'll also state my conclusions that what we do in the next fifty years will be too little and too late to "beat" it.

4 THE SOLUTION AND THE OBSTACLES

"Houston, we have a problem."

**Movie version of Apollo 13 astronauts
reporting an explosion on their spacecraft, 1995**

So, here it is in a nutshell.

Global warming is real and human activities are causing it. Worldwide, **we're releasing forty to fifty billion tons of CO_2 into the atmosphere, annually.** The added CO_2 in the atmosphere is causing a greater fraction of the sun's energy to be absorbed and retained by the earth. This added energy is slowly cooking our planet.

Halting the human-induced warming of the planet, i.e., bringing the earth's energy flow back into balance, will require a near-total transformation of the world's energy economy in a little less than three decades. **Specifically, in each of four areas -- energy, economics, technology, and international relations -- the following things need to happen:**

1. **Energy Changes --** We need to leave in the ground virtually all of what remains of nature's sequestered carbon.
    1.1. Electricity Generation
        1.1.1. We have to stop generating electricity by burning fossil fuels. That means no coal-fired electric generating stations, no natural-gas-burning power plants, and no oil- or diesel-burning power plants (other than emergency generators). We'll have to replace virtually all those power plants with solar, wind, hydroelectric, nuclear, biomass and geothermal generating capacity.
        1.1.2. Utility companies will have to install additional electric generating and energy storage capacity to serve the new

electrical demand created by the conversion of processes based on fossil-fuel combustion to electric alternatives.

1.1.3.To maintain system reliability, the electric utility industry will have to install massive quantities of electric energy storage capacity, primarily electrochemical batteries and pumped hydroelectric capacity, sufficient additional carbon-free energy resources to charge this storage plus thousands of miles of additional high-voltage transmission lines.

1.1.4.Industrial, commercial, and residential customers will have to adjust their operations and/or lifestyles to accommodate time-of-day energy use restrictions, including blackouts and brownouts.

1.2. Process Heat -- Industrial, commercial, and residential customers will have to abandon investments in durable, fossil-fuel-burning capital equipment (boilers, furnaces, ovens, stoves, hot water heaters, process equipment, etc.) and replace all of it with electric alternatives – but, importantly, only after the electrical grid is completely carbon-free.

1.3. Transportation – We'll have to discontinue the use of fossil fuels for virtually all transportation services. No gasoline-, diesel- or natural-gas-powered cars and trucks; no diesel trains; ideally, no petroleum-powered ships or airplanes. People and goods will have to be moved around by vehicles powered from a carbon-free grid or battery-powered vehicles recharged from carbon-free energy sources or vehicles/transports powered by synthetically manufactured gas or liquid products. If those manufactured mobile fuels contain carbon, the source of that carbon will have to be the atmosphere.

2. Economic Changes

2.1. The world will have to implement and enforce a worldwide carbon tax.

2.2. Everyone in most industrialized countries will have to pay between three and four times what they are now paying (in real terms) for their energy needs to receive the same level of convenience and reliability they currently enjoy.

2.3. As energy costs for all purposes increase, the transport of goods and people will be sharply curtailed. As a result, many national and local economies will have to develop other forms of economic commerce. Leisure travel will become more expensive and less accessible. Areas that currently depend upon tourism will have to diversify and find other sources of revenue.

2.4. Governments will have to allocate enormous resources in three distinct and competing areas of need:

    2.4.1. Disaster Recovery -- Supporting and rebuilding communities after weather-related disasters (e.g., floods, tornadoes, hurricanes, droughts, famines, forest fires, pandemics).

    2.4.2. Disaster Avoidance -- Investing in infrastructure to protect existing investments (e.g., sea walls and levees to protect farms, cities, and industrial enterprises) and relocating enterprises and populations away from the most-vulnerable areas to less-vulnerable areas.

    2.4.3. Preparing for the Future -- Rebuilding and greatly expanding the carbon-free electric utility grid infrastructure.

3. Technological Fixes

3.1. Advanced nuclear fission reactor concepts that have been on the drawing boards for years will have to be demonstrated, refined and deployed in large numbers.

3.2. Cost effective and efficient technologies to permanently extract CO_2 from the atmosphere, usually referred to as Carbon Capture and Sequestration (CCS), will have to be developed, refined, financed, and deployed.

3.3. Cost effective and efficient electrolysis technologies (splitting water into its constituent parts, hydrogen, and oxygen) will have to be developed, demonstrated and deployed.

3.4. In conjunction with all other mitigation measures, practical geoengineering, or Solar Radiation Management (SRM) techniques to reduce the amount of energy being absorbed in the atmosphere will have to be tested, developed, and if warranted, financed and deployed.

4. International Commitments

4.1. All countries (or at least all the major emitters) will have to commit to eliminating GHG emissions. The countries will have to honor their commitments over a period of decades regardless of how financially painful the process becomes.

4.2. Vested interests at the individual, local, regional, national, and international level will seek to protect their private interests by trying to derail the necessary changes. Climate activists will have to recognize and resist those efforts.

4.3. Because greenhouse gases mix freely in the atmosphere as they flow around the globe, some form of extremely strong treaty arrangements will have to be instituted to develop and enforce worldwide global warming response measures.

All these things will have to be accomplished to even just get to the condition, zero CO_2 emissions, whereby we're not making the excess heat gain by the earth any larger. This is not a casual exaggeration. **ALL of these things need to happen.** Here's why.

First, we know that some of our efforts to eliminate CO_2 emission from human activities will fall short of what needs to be done or take longer than expected. If we want to stop the increase in global warming, we'll have to make up for shortfalls in one area by reducing CO_2 emissions by greater amounts and more quickly in others. This "do-everything" approach is the global warming equivalent of **not** putting all of our eggs in one basket (i.e., not depending on one or two solutions).

The second reason we need to pursue a "do-everything" approach to reduce CO_2 emissions relates to goal setting. Somehow, "minimum requirements" always seem to morph into "goals." If some esteemed research entity calculates that renewable energy sources need to account for no less than 50% of all energy consumed, pretty soon **everybody forgets that 50% is a minimum requirement. People begin to think 50% is the goal.** Striving to reach a goal that's the **bare minimum** of what needs to be done almost guarantees failure. It's just safer for the planet to establish the ultimate goal of eliminating all fossil-fuel consumption.

Third, we need to follow a do-everything approach because we don't know which of the CO_2 reduction strategies will yield the biggest bang for the buck. Some will have hidden costs that will only be fully revealed over time and others will encounter unforeseen obstacles. The market and the real world will tell us what really works, but we won't know until we try everything in earnest.

And finally, the fourth reason for pursuing all possible approaches to eliminate CO_2 emissions is that we know there are some processes or chemical reactions from which CO_2 emissions cannot be eliminated. In Chapter 12 for example, I'll describe the process of making cement, which requires driving a molecule of CO_2 out of the chemical compound from which cement is made. As long as we continue manufacturing cement, CO_2 will be emitted. This source of CO_2 is not currently being captured and sequestered. If we can't eliminate this source of CO_2, we need to eliminate all those we can. The release of methane from thawing Arctic tundra is another example of a GHG emission we probably can't control. In other words, we may need to over-perform in all our efforts to curtail GHG emissions because there are some known and likely a number of unknown processes that will run counter to our best efforts to reduce emissions.

Zero Emissions – Just the Beginning

Because CO_2 persists in the atmosphere for hundreds of years, it's important to realize that even if we were able to somehow accomplish the impossible and decarbonize the world's entire energy economy, the excess CO_2 already in the atmosphere by the time we reach that historic zero-emissions milestone would continue to trap additional energy from our sun and continue to heat the planet. In other words, we will not only have to squeeze fossil fuels out of our daily energy diet, we'll also have to remove a large percentage of the excess CO_2 we will have already added to the atmosphere up until that point – **maybe a trillion tons of CO_2**. Recall that prior to the industrial age, the atmospheric concentration of CO_2 was roughly 280 ppm and at that level the earth wasn't apparently heating or cooling at an excessive rate. My guess is that we probably wouldn't need to get all the way down to those pre-fossil-fuels levels, but we may need to bring them back down to values in the low 300's. (Estimating what that tolerable concentration of CO_2 will fall to atmospheric physicists and climatologists. They'll have to assess the energy-adding mechanisms, energy-removing mechanisms, and the natural and man-made mechanisms to scrub CO_2 out of the atmosphere to calculate a concentration of GHGs that will allow the earth to come back to a livable global average temperature).

Consider this dietary analogy. Let's say you know a healthy, active individual whose dietary interests one day take an inexplicably sharp turn in the direction of cake. He can't get enough cake. He's consuming cake like there's no tomorrow. You can see that he's gaining weight, so you attempt an intervention. Try as you might, your friend is undeterred and finds a few more bakeries that can meet his cravings. More cake, more cake, and more cake until finally you gather enough of his friends to intervene and help him gradually reduce his consumption of cake. In time, his cravings for cake are tamed and he can celebrate with a big slice of . . . no, that's not right . . .he can celebrate with a modest salmon salad. He's back to eating right. Now he doesn't even like cake. You're happy, he's happy. End of story, right?

Not exactly. Your friend might be eating a healthy diet now which is a good thing and at least he's not making the problem worse. He is, however, still uncomfortably overweight. He can't do anything he used to do. Until he loses the excess weight, he won't be able to get back to his enjoyable, pre-cake lifestyle.

Fossil fuels are our cake and the excess CO_2 in the atmosphere is the extra weight we gained over the period of our uncontrolled consumption of . . . cake. Until we shed that extra weight (i.e., the excess CO_2 in our atmosphere), our world health will continue to decline (i.e., the earth will continue to overheat).

More Than a Moonshot

Some people have said we need a "moonshot" effort to combat climate change. I know what they're trying to communicate, but I think the analogy is wrong. The goal announced in 1961 by President Kennedy "of landing a man on the moon and return him safely to Earth [before this decade is out]" was aspirational as well as inspirational, but it wasn't a matter of survival. If we didn't make the moon landing in the decade of the 1960s, the moon was still going to be there in 1975 or 1980 or 1985.

Global warming is different. **Global warming is a matter of survival.** Rather than a "moonshot" effort to beat climate change, we need a worldwide "Manhattan Project" effort.

> *Global warming is a matter of survival. Rather than a "moonshot" effort to "beat" climate change, we need a worldwide "Manhattan Project" effort.*

Lessons From a Wartime Gamble

There are a lot of lessons from the Manhattan Project, the American effort during World War II to develop an atomic bomb, that apply to the question of how to best pursue a goal where survival hangs in the balance.

From Appendix B, About the Author, you know that I studied nuclear engineering in graduate school. Nuclear weapons weren't part of the curriculum, but the nuclear power industry grew out of physics research and the military effort to develop the atomic bomb so I read what I could to understand more of that history. General Leslie Groves's book, **Now It Can Be Told: The Story of The Manhattan Project** is certainly one of the most authoritative. I recommend it for a variety of reasons.

Spanning much of the 1930s, physicists around the world, especially in Europe, were using the newly discovered neutron particle to probe the structure of atoms. They systematically bombarded each of the elements in the periodic table with neutrons. Late in the decade they had worked their way up to the heaviest elements including uranium.

When natural uranium was bombarded with neutrons, researchers were initially baffled by what they observed. In time they realized that they had observed nuclear fission: the process by which certain uranium atoms split into two new lighter nuclei after absorbing a neutron. They also discovered that an enormous amount of energy was released in the process. Scientists recognized almost immediately that if a rapidly increasing nuclear chain reaction could be sustained, the resulting release of energy would be far greater than any known weapon.

As the world plunged into the Second World War, German, Japanese and American scientists convinced their respective governments that development of a superweapon, an atomic bomb, could determine which side would win the war. Albert Einstein, the most renowned physicist in the world, was recruited to convince President Roosevelt to authorize a top-secret project to develop the atomic bomb. Einstein was never directly involved in the subsequent development effort, but his endorsement was critical to the decision to pursue it. The project was given the code name: the Manhattan Project. From the beginning, it was clear that developing the atomic bomb was going to be an enormously expensive proposition with no guarantee of success. This was a time when the "home front" was growing victory gardens and donating pots and pans to be melted down to make armaments. Diverting huge sums of money away from the conventional war effort to a long-term secret physics research and bomb development effort that had no guarantee of success would have been highly controversial had it been made public. It wasn't.

Whole towns were literally built from the ground up to support the enterprise. Top scientists were diverted from other projects and secretly moved to towns that weren't on any maps. Elite teams labored in complete secrecy on basic nuclear physics principles and high energy chemical explosive technology.

Failure was not an option. Every conceivable path to success was pursued. No expense was too great. Eventually, the effort focused on two parallel pathways to the bomb. Each involved developing basic science from scratch and then applying that knowledge to building enormous, first-of-a-kind, industrial plants to produce quantities of "bomb material" measured in just tens of pounds. No one knew if either, neither or both pathways would be successful. (Ultimately, both were).

So, how is the Manhattan Project relevant to global warming? The answer is: **survival**. With everything on the line, the Manhattan Project organized teams of the most outstanding researchers, focused them on the ultimate goal, had them work tirelessly, spared no expense, and pursued multiple paths to maximize their chances of success.

In order to have any chance of slowing global warming, we need a worldwide global warming Manhattan Project: we need to act fast, engage the most talented people, spare no expense, have a public that's willing to make sacrifices and pursue every possible avenue to eliminate GHG emissions. We need to dramatically accelerate the deployment of the best technologies we currently have and expand the viable options, just as was done on the Manhattan Project.

The Challenge

The nature and magnitude of the challenge to combat global warming by eliminating CO_2 emissions are made clear in in Exhibit 4.1. (The color graphic of this exhibit is available at: https://flowcharts.llnl.gov/commodities/energy. Select the year "2019" to see the original).

Exhibit 4.1

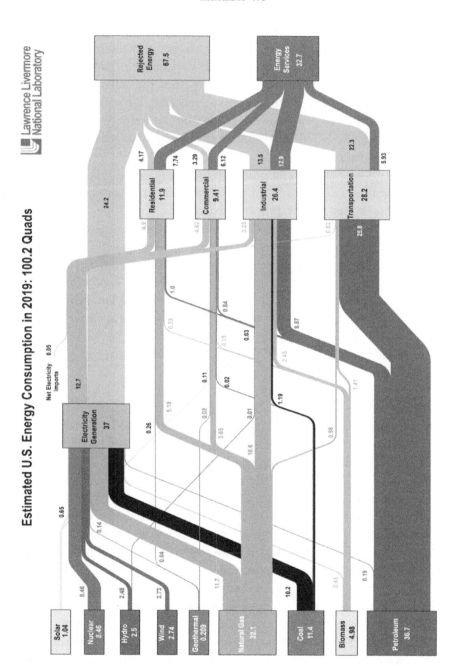

Exhibit 4.1 may be a graphic that only an engineer could love. As complicated as it appears, this exhibit is just an accounting of where all of the energy consumed in the U.S. "comes from" and "goes to." (It's sort of like your bank account: money flowing in from different sources and money flowing out to pay different expenses). Energy is expressed in quads, quadrillion BTUs. For the scientifically oriented, that's 1×10^{15} BTUs. For those less interested in numbers, just know that 1 quad is an enormous amount of energy. One quad is approximately equal to 1% of all the energy that we consume for all purposes in the U.S. in one year.

There are several points to highlight in this graphic.

1. The boxes on the left are the different energy sources. The size of these boxes is **not** proportional to the contribution each energy source makes to overall energy consumption in the U.S., but the width of the horizontal bar coming out from each box is. Plotted in this manner, it's easy to see that petroleum, coal, natural gas, and nuclear power are the primary energy sources in the U.S. totaling almost 89% of the retail energy consumed. The two horizontal bars emerging from the natural gas box illustrate that approximately 36% of the natural gas consumed in the U.S. is burned to generate electricity and most of the rest is distributed directly to industrial, commercial and residential customers.

2. The bars going into the left side of the box titled "Electricity Generation" show the sources of energy used to generate retail electricity. The bars coming out of the right side of that box show where that electrical energy is consumed. It also shows that roughly 66% (24.2 Quads) of the energy consumed in the thermal energy conversion process (i.e., turning heat into electricity) is rejected to the environment. This rejected energy is often referred to as "waste heat." While it's true this energy stream doesn't perform any useful work (most of the time), it's not exactly wasted; it's just a byproduct of the thermal energy conversion process.

   Most of the major sources of electricity come from a thermal conversion cycle. The energy released by the combustion of fossil fuel, for example, is transferred to water to make high pressure steam. The steam is directed through a turbine to turn the shaft of a turbine-generator. This process typically has a thermal efficiency of around 34% (34% of the energy released from the source fuel is converted into electrical energy). Some electric energy production processes, like hydroelectric, wind and solar, don't use a thermal cycle; instead, they produce electric energy directly, but even these carbon-free generating processes aren't 100% efficient.

3. The electric generation portion in Exhibit 4.1 deals only with retail electric sales. It excludes self-generation by residential customers.

4. The box representing residential energy consumption shows that these customers consumed 11.9 Quads of energy of all types and that roughly 65% of it provides some beneficial end use. The other 35% was rejected as waste heat at the point of consumption. A similar breakdown is also shown for the other major energy-consuming customer classifications, commercial and industrial.

5. Of the major carbon-free energy sources used to produce retail electricity (nuclear, wind, hydroelectric and utility-sourced solar) 57% was from nuclear, 19% was from wind, 17% was from hydroelectric and 7% was utility-solar.

Exhibits 3.4 and Exhibit 4.1 are two of the most important exhibits in this book. If you're concerned about global warming (which you no doubt are since you're reading this book) Exhibit 4.1 is "screaming" at us. It's telling us the magnitude and the source of the problem and where we need to direct our fossil-fuel emissions reduction efforts. I'll refer back to this energy flow graphic, and the numbers in it several times. (Note: There are additional sources of GHGs other than those associated with energy production and use. Some of them are discussed in Chapter 12).

Exhibit 4.1 is for the U.S. alone. In the coming years, all economies around the globe will have to replace virtually all of the energy currently supplied by coal, natural gas, and petroleum with energy supplied by solar, wind, hydroelectric and nuclear capacity, (mostly in the form of electricity). In addition, all future increases in energy consumption, due to population growth, improvements in the standard of living (especially in the developing world) and other new energy demands will also have to be supplied by carbon-free resources. In addition, over time, as with every other capital investment, existing "renewable" energy installations will have to be replaced with new renewables when the older ones reach their end-of-life or their performance degrades.

Exhibit 4.1 also illustrates why focusing solely on electric generation isn't enough to "beat" global warming. Retail electric energy consumption across all end-use sectors accounts for only about 20% of total U.S. energy consumption. In other words, even if we deploy enough solar panels, wind turbines, storage capacity and other carbon-free sources to supply the country's retail electric energy demand today, that would still leave roughly 80% of end-use energy coming from fossil fuels. To achieve a carbon-free energy future, we'll have to attack fossil-fuel consumption in virtually every aspect of human activity.

> *For a totally carbon-free energy future, we'll have to attack fossil fuel consumption in virtually every aspect of human activity.*

Many of the things I list as necessary to "beat" global warming will happen to one degree or another. I need to repeat that. ***Many of the things necessary to combat global warming will happen to one degree or another using a variety of technologies and strategies.*** One of those things, e.g., deploying significant quantities of renewable energy resources, is well underway already, worldwide. But, as I will demonstrate in subsequent chapters, we are not doing nearly enough nor are we doing it quickly enough to "beat" global warming.

This Is a Problem . . .

We (the U.S. in particular, and the world as a whole) are **not** on track to accomplish **even a single one of the requirements** I've characterized as essential to "beating" global warming to the degree necessary and in the time frame required.

This is an international problem, but there's no worldwide political infrastructure in existence or on the horizon capable of developing, implementing, and enforcing a plan to combat global warming.

This is an economic problem. Achieving the goal of zero carbon emissions or even net zero carbon emissions would be so expensive and so disruptive to every aspect of commerce that the global population would reject the measures that would have to be taken.

This is a reliability problem. Vast quantities of solar and wind generation resources will be deployed, but without sufficient bulk energy storage capacity, we won't be able to achieve grid reliability sufficient to break free of fossil fuels.

This is an efficiency problem. Many of the end-uses of burning fossil fuels are more efficient than their electric alternatives. Converting a fossil-fuel end use to an electric alternative, could, in many cases, increase CO_2 emissions unless the generating grid is 100% carbon-free.

This is a domestic political problem. Addressing global warming would require unprecedented cooperation and perseverance across party lines. That's not going to happen.

This is a cultural and values problem. It will require science to be accepted and to prevail over all other perspectives. Also not happening.

This is a proven technologies problem. It takes time to develop, refine, test, further refine, ramp up the production of and deploy new technologies on a scale sufficient to transform the world's energy economy. Because we have so little time to act, we can only really count on technologies that are viable and deployable today.

This is a repowering problem. The transportation sector is going electric, but until sufficient carbon-free electric generating capacity is installed to serve this new electric demand, the net CO_2 emissions reductions from the conversion to electric alternatives will be only about 20% less than what could be achieved if all petroleum-powered cars and light trucks got 45 miles per gallon.

This is a knowledge and acceptance problem. Nuclear power could be the perfect carbon-free energy resource to complement wind and solar resources. The average person, however, doesn't know very much about this technology but nevertheless has a strongly negative reaction to it.

There seems to be a widespread belief that humanity will somehow avoid this global warming calamity. This overconfidence contributes to complacency which is allowing precious time to slip away. Too many people, even those who are deeply concerned about the threat of global warming, believe that solar panels, wind turbines and EVs will carry us forward or that some marvelous new technological fix will come along just in the nick of time.

Fossil fuels are to our energy portfolio what spirits are to an alcoholic. The first step alcoholics need to take on their path to recovery is to admit that they have a problem. Isn't that why introductions at AA meetings always begin with the familiar, "Hello, I'm so-and-so; I'm an alcoholic." In order to make progress, alcoholics have to face up to their addiction.

This book is about us facing up to our fossil-fuel addiction. The first step in our "recovery" program is to fully embrace a simple truth: we're not going to "beat" global warming and we're not going to escape the ravages of climate change. Until we accept this dismaying inevitability, we will waste (even more) time, squander (even more) resources on ineffective countermeasures and most importantly, reject solutions that could actually help mitigate the problem for future generations.

The sooner we abandon the myth that we can "beat" global warming, the sooner we can begin pursuing all the necessary long-term adaptation measures concurrently with all our efforts to drive down CO₂ emissions.

> *Until we understand and accept that we're not going to escape the ravages of global warming, we're likely to waste (even more) time, squander (even more) resources on ineffective countermeasures and reject solutions that could actually help mitigate the problem for future generations.*

A lot will be done in the U.S. and around the world to reduce CO_2 emissions, but even halving per capita CO_2 emissions over the next thirty years, an extraordinarily ambitious target, will still yield CO_2 concentrations in the atmosphere that could top 500 ppm in the 2050s.

These observations, projections, conjectures, however you characterize them, are unpleasant, but I'm tired of advocates for one "solution" or another glossing over the real-world obstacles of what it would take to transition away from fossil fuels. I am certainly **not** the first one to reach this conclusion, but I want to be absolutely unequivocal about this: **we're on a glide-path to disaster. Forget about asteroids crashing into the earth. We're on a collision course with ourselves, and we're not making the mid-course corrections necessary to avoid a very unpleasant outcome.**

The best we can hope for from this point forward is that everything we do to deploy enormous quantities of wind and solar power resources, EVs and all the rest will slow the buildup of CO_2 in the atmosphere, buy us time and perhaps delay the onset of the worst consequences of global warming. But make no mistake about it: those "worst consequences" are coming.

We're already experiencing the effects of global warming. If excess heating is not curtailed, it doesn't take a rocket scientist (or a nuclear engineer) to extrapolate to how those consequences will unfold in the future.

We're not going to "beat" global warming, but I also want to be clear that **this is not a "doomsday" book.** Climate change, caused by

> *Climate change --*
> *__caused by__ global warming --*
> *__caused by__ excess atmospheric CO_2 --*
> *__caused by__ human activity --*
> *will not bring an end to human history,*
> *but it will upend life as we've known it.*

global warming, caused by excess atmospheric CO_2, caused by human activity will **not** bring a close to human history. It will **not** be the end of all life on earth. But **it will completely upend life as we've known it** during the fossil-fuel era and could ultimately cost the lives of untold numbers of people around the globe. I think it's becoming clear to most people that our future will be hotter than our past, but none of us may be able to fully comprehend just how radically different life in 2100 may be. As a result of global warming, by that date, there ~~may be~~ (strike that), there <u>will be</u> fewer people walking the face of the planet than there are today.

Pushing the Limits

When we blow up a balloon, we force air inside the surface of the balloon. As the balloon inflates the skin of the balloon experiences greater and greater stresses. If we don't stop pumping air into the balloon, eventually the balloon will burst. That's what's happening right now with the earth's energy balance. More energy is being absorbed and the earth is experiencing greater and greater stresses.

These stresses are governed by the laws of physics and chemistry. That's why this crisis is different from other dire warnings. This is not another case of crying "wolf." If we don't stop adding heat trapping gases to the atmosphere, ever-increasing amounts of energy will be absorbed in our atmosphere and earth will continue to heat up at an increasing rate.

For thousands of years, our species has flourished on this tiny little speck in the vast emptiness of space. With each passing day we are learning just how unique and intricately balanced our little "blue oasis" really is. We know what we need to do to preserve this special little place. We even have most of the technology we need to do it, but we're not doing it and we won't do it. In the span of three or four generations since 1900, we will have so dramatically altered the heat balance of our planet that we will make our earth a much less hospitable home. It will be our -- **Blue Oasis No More**.

Before moving on, I need to repeat: **I wish the conclusions I've reached were not so.** Would I want to live in a decarbonized world? **Absolutely.** Would I be thrilled to pay three or four times as much as I do now (in real terms) for energy in that decarbonized world? **No, but I'd do it – especially if the alternative was an unbearably hot world.**

I also need to repeat: this book is **not** an argument that we should throw up our hands and do nothing. On the contrary, I hope it provides a better understanding of what needs to be done on a variety of fronts and helps guide the actions of those of us who care about how life on Earth unfolds over the next fifty to 500 years and beyond.

5 WHY THIS MATTERS

**"A body in motion will stay in motion and
a body at rest will remain at rest,
*unless acted upon by some
outside force.*"**

**Newton's First Law of Motion
(Paraphrased), 1687**

Thus far, I've told you why I wrote this book (i.e., because nobody else stepped up). I've included a brief discussion about the science of global warming and I've outlined what needs to be done to at least slow down the process. I've asserted that we, the world, are not on a path to reducing CO_2 emissions anywhere near quickly enough to avoid the most severe consequences of global warming. I've presented those conclusions up front without any of the supporting information because I wanted you to know exactly where I'm going with my arguments.

Having given you this heads-up as to what's coming, I should launch into my supporting analysis. Before doing that, however, I want to tell you why I think the conclusions I've come to are important. The simple answer:

> *Knowledge is power.*

This was true long before Sir Francis Bacon made this observation in 1597. Knowledge makes it possible for us to make better decisions and impact the progression of events. The battlefield commander is powerless without reports from the front; the CEO is flying blind without information about his customers and the economy; the parent can't intervene to protect his or her child unless they know where they are, who they're with and what they're up to. The earlier we have this knowledge, the more time we'll have to guide outcomes consistent with our values and interests. Knowledge is power.

What If You Knew the Future?

What if you knew the future with certainty? What would you do with that knowledge?

Let's say that I came to you in 2008 and told you that I knew, with **absolute certainty**, that a worldwide financial collapse was imminent. And let's say that you believed me with the same level of certainty that you know that if you drop an apple from your hand, it will fall down and not up -- not a hunch; not a high probability; an absolute certainty, period, end of story. What would you have done?

Besides running around with your hair on fire warning everyone what was coming, you might have liquidated all your stock, sold all your property, moved into a rental house, put all your proceeds under your mattress and awaited the collapse. (OK, I'm sure that temporarily storing your life's savings in cash under your mattress **wouldn't** have been the "best" strategy, but this is not a book about investing). Anyway, you get my point: you would have **used this knowledge** to avoid guaranteed, short-term losses and make investments that would secure your financial future.

In fact, some people did see the coming of the 2008 financial collapse **with absolute certainty.** Armed with this knowledge, they avoided a lot of losses and made a lot of money. They didn't know exactly when the collapse would happen, but the warning signs flashed brighter and brighter as that day came nearer. They knew that the economy was way out of balance, **precariously and unsustainably out of balance**. They knew it was certain to implode, sooner rather than later --and that's exactly what happened.

The immediate consequences of the financial collapse were dramatic, and the initial decline played out over a period of months. The consequences of global warming will not be like that. The consequences will be gradual, but relentless. They will progress from bad (i.e., today) to devastating, and from devastating to catastrophic over a period of decades.

As I tried to make clear in the Introduction and in the previous chapter, the conclusion that we will not "beat" global warming is not an excuse to throw up our hands in despair and do nothing. I'm urging quite the opposite.

To this end, I have four goals:
1.   I hope this book provides those of you who've been sitting on the side lines thinking "this will all work out," with a deeper understanding of the challenges of global warming and the

urgency of the issue. Armed with this knowledge, I hope you'll make changes in your own lives and demand more concrete and comprehensive actions by the institutions (governmental, corporate, religious) in your lives. I especially hope that this book reaches young people, who are, by all accounts (and not surprisingly), more concerned about global warming because **it's THEIR world that hangs in the balance**. I hope that the perspective they learn in this book will better equip them to persuade their parents and friends that they too need to make combating global warming **the pivotal issue** for determining which politicians win their support.

2. I hope this book helps those of you who have narrowly focused on a single dimension of the problem that you need to broaden your field of engagement and redouble your efforts to address global warming in the broadest, most comprehensive way.

3. I hope this book demonstrates why **all viable options need to be part of the solution, including some that may not be appealing to you**. And lastly,

4. I hope it helps move energy issues and emissions concerns to the forefront of everything you do.

The stakes couldn't be higher.

Because the increase in greenhouse gases in the atmosphere is effectively cumulative, it's vital that we do everything we can, as quickly as possible, to cut CO_2 emissions. In addition to doing everything we can to delay and diminish the onset of the worst consequences of global warming, we also need to implement policies that will help us transition with less pain to the inhospitable future that awaits us. How that plays out for you is partly a function of where you are on the spectrum of opinions about global warming and your role in the overall social fabric.

If You're a . . .

If you're a "climate activist," I hope this book makes clear that your goals need to be far more encompassing than simply putting solar panels on every rooftop and an EV in every garage. These are certainly good, visible steps forward, but it's going to take a lot more to "beat" global warming or even just slow the rate of warming of the planet. The goal you are pursuing is nothing less than restructuring the worldwide energy economy. Solar panels, wind turbines and EVs alone won't make that happen. Work to reduce all GHGs and as you do, recognize that you will be confronting powerful countervailing forces attempting to maintain the status quo.

If you're an "environmentalist" who believes that wind and solar are the only answers, maybe this book will encourage you to rethink your opposition to nuclear and hydroelectric power plants (where they make sense). The current, more advanced generation of nuclear designs almost certainly address many of the concerns you have. Learn what's possible. **Delaying or preventing nuclear power from becoming part of the future energy mix will make the global warming crisis worse.**

If you're a "voter," don't accept at face value the promises of candidates who declare that your city or state will be carbon-neutral by 2040. First, those politicians aren't going to be in office to be held accountable in 2040. Second, **long-term goals without near-term milestones are meaningless**. Insist that politicians lay out credible plans with intermediate yearly goals so that progress toward the ultimate objective can be measured. And third, pin down those politicians on what they mean by carbon-neutral or net zero. Are they referring to the electric grid of today or the electric grid of the future that will need to supply electric energy for all those new energy end uses that currently use fossil fuels? Most of the time they're referring only to the existing electrical grid.

If they tell you they're striving to achieve **net zero emissions** (i.e., additions of man-made GHGs equal man-made GHGs removals), be aware that they're not talking about cutting actual emissions to zero. They're using credits, often of questionable effectiveness, to offset emissions they can't or don't want to eliminate because doing so might have painful political or financial consequences.

One way companies and governments can earn emissions reduction credits is to plant trees because trees extract CO_2 from the atmosphere. I love trees and support tree-planting almost anywhere, but I seriously question whether tree-planting should be used to earn credits against current emission. Trees and a large fraction of the carbon they contain can literally go up in smoke, returning at least that part of the carbon in them that's above ground to the atmosphere. We all watched in horror in the summer of 2021, when fires destroyed great swaths of forests, not only in the U.S. but all around the world, in a matter of weeks or months. Rather than pursuing remedies that might turn out to have only short-lived environmental benefits, carbon credits, to the extent they're used at all, should derive from more enduring changes in hardware, appliances, durable equipment, and process changes. I'd much rather see actual emissions from human activities be driven down to zero without relying on so-called "carbon credits." I'd much rather treat tree planting as an investment to draw down excess CO_2 concentrations in the atmosphere after we get to "real-zero."

If you're a "politician," don't make promises you can't keep. Do your homework. Know what you're talking about; use your office as a public servant to inform and educate. Respect the intelligence of the electorate. Making promises that never come to fruition can make the problem worse. A complacent electorate wants to believe this problem can be solved and that it won't be painful to them personally. Perpetuating those beliefs contributes to inaction and squanders precious time that we don't have.

If you're a "planner," understand the consequences of global warming and the extreme weather it will cause. Do your planning with the understanding that the worst-case climate scenarios are ahead. Implement policies that achieve long-term sustainability objectives such as discouraging or preventing new development in areas vulnerable to fires or floods. Similarly, implement policies that gradually phase out existing developments in vulnerable areas over an extended period of time, but begin that process now. This won't be popular, but it may prove to be more palatable, safer, and cheaper than recovering bodies from devastated communities after future storms that make Superstorm Sandy look like a cakewalk. .

If you're a "regulator," ensure that the regulations you promulgate move people's choices in the direction of sustainability and environmental balance, even if those choices are not the least-cost solution.

If you're a "global social justice activist," recognize that you're in for tough challenges in the decades ahead. In addition to all the obstacles you normally face to advance your causes, you'll also be battling the effects of global warming. Nation-states will fail and large numbers of people will become economic and climate refugees.

If you're a "military or State Department planner" assessing strategic threats, don't assume efforts designed to curb global warming will be successful. Plan for the worst.

If you're a "skeptic" who isn't convinced that climate change will be as bad as it's being portrayed, allow for the fact that **you might be wrong**. You don't expect to die young in an accident, but to protect your family you might still buy term life insurance policies on both parents . . . just in case. If you care about the future of the planet, maybe you should support decarbonization efforts . . . just in case you're wrong and the scientists are right.

If you're a "homeowner/consumer," make long-term energy efficiency a foremost consideration in your durable goods purchases. Don't make decisions based on a one- or two-year time horizon. Think long-term. If you own a home, consider installing a solar panel system and when the time is right, an energy storage battery system; if you're a renter, ask your landlord to consider solar panels. If you buy a car, buy an EV or a vehicle that gets 45 mpg or more. Only buy stuff that can be repaired. Even better, don't buy as much "stuff," period.

If you're an "appliance or durable goods manufacturer," recognize that your customers are going to be increasingly interested in energy efficiency, durability and products that may be easily and inexpensively repaired.

If you're a "parent," prepare your kids to compete in an increasingly difficult and challenging physical environment, one in which their quality of life will almost certainly **not** be as favorable as yours or what they enjoyed growing up.

If you're "religious" and revere all God's creations as sacred, don't sit in the last pew. Do everything in your power to at least slow down the destruction of God's green earth and everything you hold dear.

If you're a "diplomat," act urgently to facilitate and enforce international cooperation. Begin planning today for a world that will have to accommodate countless climate refugees and regional disasters including some that may not be able to be alleviated.

If you're "busy fixing" one of society's other ills, i.e., public access for handicapped individuals, income inequality, pet rescues, curing cancer, [insert here the name of the cause about which you are most passionate], you need to divide your attention and have at least two "causes." Whatever else you're trying to "fix" won't make any difference if we continue to "trash" the planet.

If you're the "nuclear industry," get your new reactors designed and components tested. Put a premium on concepts that are passively safe, as simple as they can be, minimize the financial risk to owner/operators and can be deployed quickly. Endorse the reprocessing of spent fuel either domestically or through an international agreement. Lobby to make a high-level radioactive waste storage facility operational. Ensure that you have a persuasive, honest story to tell a skeptical public. Start educating the public now.

If you're . . . I could go on.

If you truly understand that we're not going to "beat" global warming, wouldn't it make sense to do everything you could now to at least delay as far into the future as possible the time when those truly onerous conditions will be upon us? Wouldn't it make sense to complement your GHG-reduction activism by supporting other proactive measures to prepare for a much different and decidedly less comfortable future? Wouldn't it make sense to reconsider all your assumptions about energy policy, land use, the role of government, etc. to ensure that what you're advocating is moving us along the decarbonization path as quickly as possible and preparing us for a much less forgiving future?

Mitigating Losses: One Example

Let's say you've accepted that we're not going to "beat" global warming. Let's say you understand that our experience of more devastating wildfires, and more frequent and severe flooding, hurricanes and tornadoes, aren't random, one-hundred-year anomalies. With this knowledge, how might you, the nation and the world prepare for that future?

For one thing, you'd want to discourage or even prevent land development in areas prone to damage or destruction from extreme weather events. At the very least, you'd want to discontinue the practice of helping people rebuild their homes and their lives in these vulnerable areas after a disaster – what economists call a "moral hazard." If the nation adopted a policy that shifted 100% of the risk to property owners, at least the rest of the nation wouldn't have to pay for someone else's choice to live in an area vulnerable to fire or flood. Over time, rising casualty insurance premiums would depopulate the most susceptible areas, all without government interference or expense. The process would also save lives. Depopulating high-risk areas might take decades, but knowing what's coming our way, we could begin now and make that transition less painful and less disruptive. This is just one example of how knowledge, if we use it, could help ease us into an unforgiving future.

There are few aspects of our current lives that will be untouched by the changes global warming will bring. Aside from trying to delay and lessen the severity of what's coming, we have to figure out what those changes will be and how we need to prepare.

A Call to Action

If, after reading this book, you agree with my assessment, I hope you'll feel some urgency to do something more and something different than what you may have been doing. If you previously thought we would somehow muddle through this crisis (as we have with so many other crises) and finally understand that we're not going to "beat" global warming, then I will have done at least part of what I set out to do. With this knowledge, the next steps will be up to you.

Throughout this book I suggest actions we need to take. These recommendations are certainly not an exhaustive list. Many of you have specialized knowledge that will allow you to identify those things, big and small, that serve two goals: to help bring our earth back into (energy) balance in the near term, and to ease the transition into a difficult future in the long-term.

[*Input] -- Action Items: Please supplement my list of policies programs, and actions (Chapter 18), we need to take in order to slow global warming. Provide your suggestions through the BlueOasisNoMore.com website for consideration to be included in the second edition. [*Input - End]

Preparing For the Future

Confronting climate change is a bit like preparing for retirement. We know with certainty that retirement is coming, so we prepare. We act. In anticipation of those future retirement years, we consult experts who can advise us on how to grow a retirement fund. It's a process that we know depends only on us and will extend over decades. We know that the earlier we start, the more successful we're likely to be. Postponing and then attempting to do it in a short time can be painful and has a much lower probability of success. We know we'll have to make sacrifices now to make our future retirement dreams a reality. At the very least, we don't want to be a burden on our children. If we're successful, a financially secure retirement is the ultimate payoff.

Acting to delay and lessen the impacts of global warming has a lot of parallels to preparing for retirement. We rely on experts to tell us what we need to do. We know it'll take decades to do what needs to be done. We also know that the longer we delay, the harder it'll be to reach our goal and that our chances of success will be lower. We know that we'll have to make sacrifices now for a better long-term outcome. At the very least, we want to minimize the burden of our emissions today on future generations.

Those are the similarities, but there are also notable differences. The rewards for saving for retirement are personal and depend only on our actions. The rewards for lessening the impacts of global warming are distant, remote and depend on all of us, all of our (world) neighbors.

What we do between now and 2050 or 2075 will dramatically impact the next 500 years on earth. While the immediate unpleasant consequences of global warming can't be avoided, we may be able to change the trajectory just

> **What we do between now and 2050 or 2075 will dramatically impact the next 500 years on earth.**

enough to make a significant difference for future generations. **That's why this matters.**

Not What I Wanted to Hear

Some folks reading this book may criticize it as a downer. They may reject my conclusion because it's an unpleasant, unsettling message.

To those critics, I pose this question: if someone wrote a book documenting racism in our society with all of its ugly, damaging, and painful consequences (there are many such books by the way), would you reject such a book because its content was unpleasant? I don't think so. If it explained the real-world impediments to ridding the world of racism, would you despair and give up? I hope not. Ideally, such a book might help you focus your efforts on the areas where you might affect the greatest change. It might strengthen your resolve to fix the problem.

In the same way, I hope the unsettling conclusions in this book only strengthen your resolve to do everything you can to fix it, even if "fixing it" in this case only means slowing the onset or lessening the impact of global warming on future generations. I can't control how you react to the message I'm delivering. I only hope it fills you with a sense of urgency and energy, and that the list of "Action Items" gives you some direction.

Others may criticize me personally as a defeatist. I'd ask them to consider the sophomore materials engineering student watching the World Trade Center towers burning on 9/11. The future-engineer would have known that the yield strength of the steel used to construct the towers was decreasing minute-by-minute as the raging fires drove the temperature of the metal beams higher and higher. It would have been clear that the ability of the steel beams to support the weight of the structures above the fires would soon be exceeded. Knowing that the buildings were going to collapse, the young engineer would have been screaming at the TV for the firemen to turn around and get out. Even if they could have heard those screams, their firefighter-DNA would have made them deaf to those frantic pleas.

Would you characterize this budding young engineer as a "defeatist" for using his or her knowledge to try to lessen the consequences of a disaster that was imminent?

That said, I'm certain I'll still be roundly criticized by some as a "defeatist," but I'm also certain (or at least hopeful) that many who read what I've written will realize that **we really need to do something dramatically different than what we have been doing.**

Finally, moving on to the supporting analysis. In the remainder of the book, I'll discuss each of the requirements I laid out in Chapter 4. Because attacking global warming means "electrifying" virtually every energy-consuming activity with carbon-free resources, I'll begin with an analysis of what it would take to decarbonize our existing electrical grid.

6 ELECTRIC POWER: GENERATION AND USAGE

"Power to the People!"

1960's Cultural/Protest Expression

Simply stated, the solution to global warming is to turn off the fossil-fuel spigot and then draw down the excess CO_2 in our atmosphere that's trapping more of the sun's energy. So, the first step is to stop pumping CO_2 and other GHGs into the atmosphere. The solution is that simple . . . **and that hard**.

There are many sources of greenhouse gas emissions. Exhibit 6.1 is an EPA estimate of the percentages of GHG emissions from the primary fossil fuel end uses for the world and for the U.S.

Exhibit 6.1 Greenhouse Gas Emissions by Sector

Category	Global	U.S.
Electricity and Heat Production	25 %	28 %
Agriculture, Forestry, Land Use	24 %	9 %
Industry	21 %	22 %
Transportation	14 %	29 %
Buildings	6 %	12 %
Other	10 %	
Total	**100 %**	**100 %**

Exhibit 6.1 is one way of categorizing the emissions from various sectors. Other categorizations lead to slightly different percentages for the different sectors.

Rather than focusing on the differences between the various approaches, there are two primary takeaways from tables like these. The first takeaway is that GHG emissions come from a number of different energy-consuming activities. The significance of this is that attacking global warming means waging war on multiple fronts. The second takeaway is that the major sectors giving rise to GHGs are: electricity generation (coal and natural gas), transportation (petroleum), industry (coal and natural gas), agriculture (livestock operations) and direct combustion (natural gas) as a heat source in commercial and residential applications.

The primary strategy for combatting global warming is to electrify all fossil-fuel-consuming activities with electric energy generated from carbon-free resources. To discuss what that entails, this chapter reviews the structure and operation of the electric industry. This review is intended to highlight the infrastructure we need to replace and the technologies we have to make that transition. The four chapters following this one describes the options available to electrify not only the existing electrical grid but also the other energy sectors that will eventually need to be decarbonized.

Electric Utility Structure and Econ 101

The electric energy supply grid in the United States (and most developed nations) is a complex mixture of resources. Before utility deregulation in the U.S., the utilities served the electric energy demand of their customers by deploying generating equipment utilizing a variety of different fuel types. For example, a utility might operate a mixture of hydroelectric, coal and nuclear plants, its resources with the lowest marginal cost, to meet its base load (i.e., that level of energy demand on its system that's present all day, every day). Then, it would typically have additional natural-gas-burning plants capable of being ramped up and down ("load-following units") to match changing customer demand during the day and night. These load-following units and their base-loaded units plus electric energy imported from neighboring utilities over high voltage transmission lines would usually suffice to meet customer demand >99% of the time.

In addition to these resources, a utility would typically have additional generating equipment connected to the grid but operating at a low power level. Utilities have this generating capacity, called "spinning reserve," so that it's available to feed energy into the grid almost instantaneously. Spinning reserves help ensure that the utility can serve its customers' energy demand even if one of its online power plants experiences a fault that requires it to be disconnected from the grid without warning. Spinning reserve boosts the utility's reliability factor.

Sometimes, however, customer demand spikes far above normal as it might on the afternoon of the third day of a summer heat wave when customers just can't bear the heat any longer and crank up their air conditioners. On those days, the utility pushes all of its generating capacity to the limit. If it still can't satisfy demand, it's forced to operate its most expensive generating resources referred to as "peakers." These generating units are typically jet engines attached to electric generators. These peaking units may literally be used only fifty or fewer hours per year, but they are critical to maintaining system reliability at a high level.

In states where the electric utility function has been deregulated, power generating enterprises do something similar to what the conventional integrated electric utility did for years to deliver energy to the grid to meet customer demand. In the deregulated electric energy market in California today, an Independent System Operator (Cal ISO) manages the flow of energy on high-voltage transmission lines and operates a competitive wholesale electric energy market. For this discussion, however, it's simpler to describe how an integrated electric utility system works where the generation, transmission and distribution facilities are owned and operated by the utility company. In either case, all of the same principles and challenges apply regardless of the transactional structure of the energy marketplace. One of the big differences, however, at least from the perspective of a former utility employee, is the commitment to service that traditional utilities embodied compared to the profit motive that drives current suppliers.

Historically, utilities built a variety of capital assets including generating plants, a network of high voltage transmission lines, substations (to "step up" and "step down" transmission voltages), and distribution lines to its customers' homes and businesses. After making these investments, utilities would recover their costs over the useful life of the assets. As long as the utility performs competently, this investment recovery mechanism was used even if an asset was taken out of service early or underutilized compared to what had been anticipated. This is where state public utility commissions enter the picture. They oversee the utility's performance and set electricity prices to cover the utility's capital and operating costs and provide a reasonable rate of return (a profit) to the investors who provided the money to build the assets.

Leaving aside the issue of energy storage (which I'll address in Chapter 9) **electric energy is generated and used on an**

> *Electric energy is generated and used on an instantaneous basis.*

instantaneous basis. Supply must equal demand at every moment to maintain system voltage and frequency within narrow bands. The requirement to always match supply (generation) and demand (instantaneous energy usage) is a vitally important requirement in terms of understanding why we're not going to "beat" global warming. This is such an important concept, it's worth repeating.

The amount of energy delivered onto the grid at any moment has to match the amount of energy being extracted from the grid at that moment. If the amount of generated energy doesn't equal the amount of energy demand, the electrical voltage or frequency on the grid could increase or decrease, potentially harming customers' equipment.

The definition of high-quality electric service is electric energy that's provided at a specified frequency and voltage within tight tolerances and is available any time of the day, every day of the year, at the flick of a switch. When it isn't available, our lives (or at least our regular routines) fall apart.

The power that electric utilities produce and distribute to their customers is usually measured in kW (kilowatts) or MWe (megawatts electric, thousands of kilowatts). The amount of **power** a customer uses at any moment in time is also measured in kW and is typically referred to as "demand." The amount of **energy** a customer uses is determined by the amount of power their appliances or equipment demand and the length of time that equipment is operating.

In other words, energy is power consumed over a period of time. Electric energy consumption is usually measured in kilowatt-hours, kWh. Power and energy are therefore related, but they're very different. Being clear on this difference is also critical to understanding the challenge of "beating" global warming and going "green."

For example, a one-kW commercial refrigeration system operating continuously for a full day would consume 24 kWh of energy (1 kW for twenty-four hours). A 24-kW electric furnace operating for one hour would also consume 24 kWh of energy (24 kW for one hour and 0 kW for twenty-three hours). Both end uses consume the same amount of energy, but these two power demands on the electric grid are entirely different.

Central to Our Lives

Maintaining a **reliable** electric grid (24/7 availability at specified voltages and frequencies) is the single most important requirement impacting our ability to "beat" global warming. It's also the most difficult. In developed countries, reliable high-quality electric energy is taken for granted by consumers. It's easy to forget that achieving electric system reliability >99+ is truly a modern miracle (as are: jet travel, clean water systems, medical treatments, sewage and waste disposal systems, and many other things we take for granted). We design our lives and processes around the expectation of reliability. That's not the case in many parts of the world where people design their lives and enterprises around intermittent, scarce, and often expensive electric service.

After the catastrophic 2018 Camp Fire incinerated nearly the entire town of Paradise, CA, the utility serving the area was forced to adopt a strategy of pre-emptively cutting service (Public Safety Power Shutoffs, PSPS) to certain locations during severe wind events that could spark uncontrollable forest fires. Those outages have been controversial, and frankly, I don't know enough to offer an opinion on whether or not those PSPS were warranted. I do know that the interruption of electric service to those communities demonstrates just how dependent we've become on high quality electric service: businesses couldn't open, people couldn't go to work, homes and streetlights went dark, the contents of refrigerators spoiled, home medical equipment stopped operating, etc.

The Great Texas Freeze of February 2021 is another recent reminder of just how important reliable electric service is in modern society and how services of all kinds that we take for granted are intertwined.

The mindset of an electric utility employee is that the lights should never go out. More than anything else, the system must be designed and operated to ensure that electric service will be available 24/7 under virtually all weather, economic and other conditions.

Daily Load/Demand Profiles

Exhibit 6.2 shows the demand for power from the electric utility grid in Southern California on a typical summer day and a typical winter day in 1988.

Exhibit 6.2

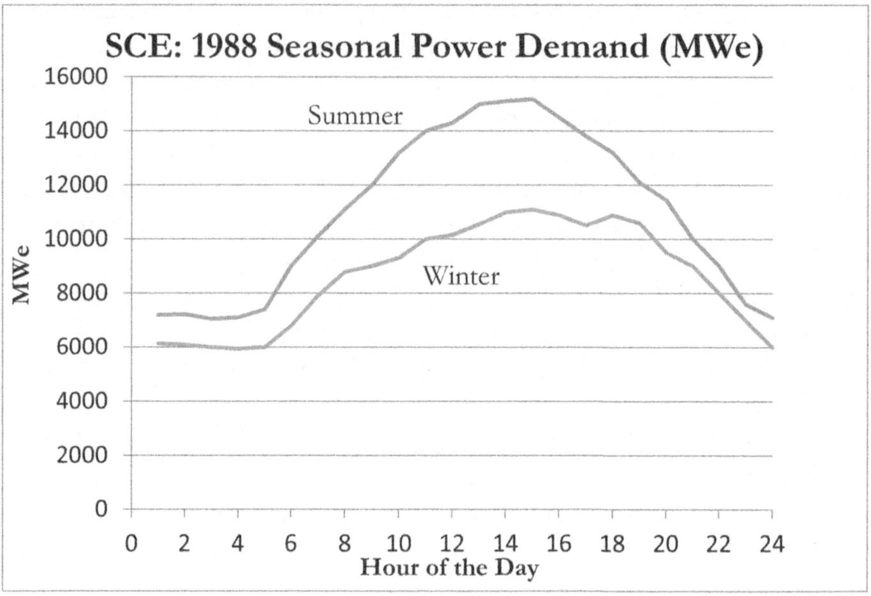

SCE: 1988 Seasonal Power Demand (MWe)

These demand curves were typical for my former employer, Southern California Edison (SCE), in 1988. I've intentionally used these **thirty-year-old data** for several reasons.

First, in the 1980s, SCE was a leader among electric utilities in developing and promoting renewable energy resources and electric vehicles. I compiled these data for a study identifying what our utility would need to do to accommodate the introduction of massive numbers of EVs recharging on the SCE system. As it turned out, our study was three decades too early, but the fact that we performed it at all illustrates the mindset of a utility to look far into the future to anticipate its customers' needs. In addition to being done decades before the EV energy demand materialized, that study included certain implicit assumptions about utility generating resources. Those assumptions seemed rock solid at the time but are no longer valid for an electric grid in 2022 that includes massive quantities of intermittent renewable resources. Consequently, many of the study's recommendations would be entirely different if that same study were performed today.

The second reason for displaying these **old data** is to highlight the fact that in 1988, the electric utility provided virtually all of the electric energy used by its customers. In other words, a utility customer's total power usage was essentially the same as their power demand on the electric utility grid. That's not true today in California and other areas where significant quantities of rooftop solar systems have been deployed. Today, customer-owned solar and other customer-owned generation resources reduce the demand (kW) on the electric utility grid and simultaneously reduce the utility's total energy sales (kWh) to customers.

In the wake of utility deregulation, Cal ISO manages the electric grid in California. Massive amounts of renewable resources have been added to the power mix in the state. Management of the grid has become a lot more difficult. In 1988 it was adequate to think in terms of summer and winter power demand profiles. Now, however, different times of the year present different challenges for the utility and those challenges are shifting as more renewables, predominantly residential-owned solar panels, wind turbines, and more EVs are added to the system.

The shape of the daily power demand curve (the rate of energy consumption over the course of a day) is critically important. Our power demand reflects our daily activities, the number of daylight hours, air conditioning and heating needs, etc. We wake up in the morning and start using energy. We go to work and use a lot more energy. As activities at our places of business ramp up during the day, we use even more energy. In the late afternoon, the day begins to wind down and our energy usage begins to drop. We go home in the late afternoon or early evening and use some more energy. We make dinner and use energy. We watch TV or other entertainment in the evening and use still more energy and finally we retire for the evening, at which time energy usage by most customer classes (residential, commercial, and industrial), begins to scale way back. That energy consumption pattern is reflected in Exhibit 6.2.

The Grid Doesn't Sleep

One of the things that should stand out from the demand curves in Exhibit 6.2 is that when the sun goes down, the grid doesn't shut down. There are still significant loads on the system both before sunrise and after sundown. In 1988 the night-time power demand on the Edison system was about half of the peak daytime demand. As will be discussed in Chapter 8, these night-time loads represent a significant operational problem and capital investment requirement on a grid that depends heavily on solar power and is trying to decrease its dependence on fossil fuels.

Regions Other Than Southern California

The shape of the demand curves in Exhibit 6.2 were typical of Southern California, where residential and commercial loads were dominated by air conditioning in the summer, and lighting, heating, and home activities in the winter. Other areas of the country (and the world) have their own characteristic seasonal energy usage profiles. Those profiles in Chicago, Miami and Idaho Falls are all different at various times of the year from those in Southern California. In some ways, however, they're all similar because the energy usage curves reflect how most of us live our lives, i.e., active during the day and sleeping at night. The bottom line is that electric utilities everywhere face the same energy delivery challenge: to serve the power requirements of its customers at every moment of the day with whatever resources it can muster.

While the daily and seasonal demand curves in Southern California may differ from those in other parts of the country, it's still useful to consider them in some detail. If we do what needs to be done to dramatically reduce or eliminate CO_2 emissions from electric generation, massive amounts of renewable energy resources will have to be added to the nation's electrical grid and fleets of EVs will have to be rolling down the

> *The "future" arrived in California before the rest of the country. California will serve as a laboratory for working out some of the issues associated with utility operations in a world increasingly reliant on renewable energy resources and EVs.*

nation's highways. That's happening today in California, perhaps more than in any other state. In other words, the "future" has arrived in California before it has in most of the rest of the country. California will serve as a laboratory for working out some of the issues associated with utility operations in a world moving away from fossil fuels.

Choose Any Two

I recall a lecturer who once described the challenges of project management in terms of three parameters: time, cost, and quality. He argued that you can have any two you choose, but not all three. A project can be fast and cheap, but it won't be high quality. Or, it can be low cost and high quality, but it won't be fast. Or you can have a high-quality product in a short time, but it won't be cheap.

The electric generation system can be similarly characterized by three parameters: reliability, cost, and environmental friendliness ("green-ness"). As in the project management model, you can have any two you choose, but not all three. The generating system can be reliable and cheap, but it won't be green. This might describe the U.S. utility systems through the year 2000. Or, it can be comparatively inexpensive and green, but it won't be reliable. This might describe an isolated community served by solar arrays that's not connected to a central grid. Or, it can be green and reliable, but

> *Green, Cheap and Reliable: Choose any two.*

it won't be cheap, i.e., the electric system that we need to get to in the not-too-distant future.

Utility companies go to great lengths to make their systems reliable and robust all the way from generation to energy consumption. The utility distribution system is separated into individual circuits serving different neighborhoods or businesses. These circuits can be isolated from one another. That's why, when a motorist has an "up-close and personal encounter" with a power pole, the crash may bring down the power pole, but not the entire grid. Instead, it's the homes and businesses on that circuit that are affected.

Some of a utility's generating resources may be concentrated on one site, but most are dispersed geographically to be close to the point where the energy is being used. Having multiple sites also helps ensure that plant maintenance, an accident or other localized event won't degrade the utility's ability to serve its customers. If one generating plant has to be taken off-line, others are typically available to make up the loss.

Utilities typically track the performance of their generating plants using a number of different parameters, two of which are **capacity factor** and **availability**. **Capacity factor** is a number equal to the total amount of energy generated by the power resource over a period of time (such as one year) divided by the total amount of energy the plant could have produced had it operated continuously at full power for that entire period. **Availability** is a number equal to the total number of hours the power plant was functionally able to produce power in a year (regardless of whether or not it was called upon to operate) divided by the total number of hours in a year (8,760).

The two performance parameters are often quite different. To illustrate the point, consider your car: it may have an availability (i.e., functionally able to move you from place to place) of 99%, while its capacity factor (the number of miles actually driven in a year compared to the number of miles the car could have been driven if you drove it 24/7 for a year) might be 3%. A high availability means that a plant is likely to be able to generate power when called upon to do so. A high capacity factor means that the plant actually generated as much power as it could most of the time. Utilities want all of their assets to have high availability and their generating resources with the lowest operating costs to have the highest capacity factors.

Having a variety of generating facilities, fuel types and transmission-line interconnections helps ensure that the utility can deliver the high level of system reliability we all expect and enjoy. Achieving the same level of reliability on the decarbonized grid of the future will be a monumentally difficult and expensive proposition.

Action Items

In the area of energy generation and usage, here are a few of the projects and policy initiatives we should pursue to combat global warming:

1. Phase out any remaining coal-burning electric generating plants within five years.

2. Identify sites where stream flows are sufficiently secure to build micro-hydro generating facilities. Build garbage-burning generating facilities on an accelerated time schedule.

3. Implement dynamic energy pricing nationwide within three years. It's essential that customers receive correct price signals in real time so that they can adjust their energy use profiles accordingly.

4. Develop programs to rapidly increase energy conservation. Increase residential and commercial building insulation standards. Increase appliance efficiency requirements.

5. Create and publicize a quarterly, decarbonization report card and the atmospheric concentration of GHGs. Track energy production (kWh) from: 1) all each carbon-free energy resources (i.e., utility scale solar power, residential rooftop systems, wind, hydro, garbage burning and nuclear), and 2) total energy deliveries from chemical energy storage systems and pumped hydro facilities.

7 THE CARBON-FREE "ELEPHANT" IN THE ROOM

"Snatching defeat from the jaws of victory."

**A turn-of-phrase appearing
in the New York Times
(and other papers), 1891**

Nuclear power. If ever there were two words that could strike fear in the hearts of environmentalists, these are those two words. Facing the prospects of a warming planet, however, and not surprisingly, many prominent global warming activists and others are changing their tune and embracing the technology they once considered "verboten." (Given that I have advanced degrees in nuclear engineering and spent a good bit of my career working in this field, you may not be surprised that I would devote a chapter to this technology. You may, however, be surprised by my conclusion, but maybe not for the reasons you suspect).

(Spoiler Alert: Nuclear Power is NOT going to save us.)

Nuclear power does have its virtues. Foremost among them is that it's carbon-free. Considering the urgency and importance of decarbonizing our energy sources, it's surprising that so many critics dismiss this attribute as if it weren't that important. In addition, there's an abundance of uranium and thorium, the feed materials from which to breed and fabricate fuel. Furthermore, it's proven and dependable. The U.S. derives around 17% of its electricity supply from carbon-free nuclear power plants. Over the last two decades, 70% to 75% of all the electric energy consumed in France was generated by domestically operated nuclear energy. That's not a typo: 70% to 75% from nuclear.

There's a reason France has led, and still leads, the world in the percentage of nuclear power on its national grid. That reason is "necessity." France doesn't have an abundance of fossil-fuel resources and we all know that scarcity has a way of focusing the mind. For France it was a matter of economic survival and energy security. They had to either import massive amounts of coal, oil, natural gas, and electricity or go nuclear. They chose the latter course and are today exporting carbon-free electricity to neighboring countries. In perhaps one of the most short-sighted national policies ever devised, under France's (current) Energy Transition Law, adopted in 2015, the contribution of nuclear power to the French grid is slated to **decrease** to 50% while natural gas increases from below 10% to near 30% over the next two decades.

Ramping down nuclear production and burning more natural gas to generate electricity is completely contrary to efforts to combat global warming. It will be interesting to watch French public opinion and government policy evolve as global warming and energy security issues become more prominent. The French plan seeks to reduce total energy consumption but given France's demonstrated ability to manage nuclear technology and their excellent safety record, their plan to ramp down nuclear and ramp up its reliance on natural gas consumption makes, bluntly, no sense. Instead of reducing their production of carbon-free nuclear energy they could have aggressively pursued conservation goals while **simultaneously** expanding their nuclear production. This approach would have put the French in a position to export even more carbon-free energy to its neighbors and reduce all of Europe's dependence on natural gas from Russia. I fully expect France to reverse course, again, within the next few years.

Prior to the Russian invasion of Ukraine, Germany seemed poised to accelerate its retreat from nuclear power. The war, however, has highlighted the national security imperative for Germany (and many other countries) to decrease their dependence on imported fossil fuels. It will be interesting to see how quickly and to what extent, Germany reconsiders its relationship to nuclear power. [If "Necessity is the Mother of Invention," it may also be the "Stepmother of Acceptance."]

> *If Necessity is the Mother of Invention, it may also be the Stepmother of Acceptance.*

Nuclear power constitutes about 17% of all electric energy generation in the U.S. Worldwide, nuclear power accounts for just over 10%. Coal and natural gas are used to generate six times that amount.

While nuclear power has many virtues, it's also perceived by the general public as having serious flaws. The two primary objections to nuclear power relate to public safety and radioactive waste disposal.

One accident at an American nuclear power plant and two disasters abroad are indelibly etched in the mind of the public: Three Mile Island (TMI) in Pennsylvania, Chernobyl in Ukraine, and Fukushima in Japan. I won't dwell on these accidents other than to give a brief overview of each.

Three Mile Island (TMI)

The Three Mile Island nuclear power station was about 90 miles north of Washington D.C. (I say "was" because both reactors that once operated on the site are now shutdown. Unit 1 was shutdown permanently in 2019 after operating safely and reliably for 45 years).

TMI Unit 2 was shut down and severely damaged in an accident in March 1979. The accident was a financial disaster for the utility company operating the plant, but not a public safety catastrophe. No lives were lost and no land area around the plant was rendered uninhabitable. The minor release of radioactive material as a result of the accident was mostly in the form of chemically inert radioactive gases that were quickly dispersed.

The consequences of the TMI accident, i.e., no deaths and a negligible release of radioactive isotopes, were not just a lucky break. Civilian nuclear power plants in the West are designed and constructed to withstand a wide range of accidents without damage to the plant or adverse public health consequences. They incorporate layers of backup systems to control the reactor, keep the core adequately cooled and prevent the release of harmful radioactive material. The reactors are designed to withstand a wide range of design-basis accidents, one of the most severe of which is a loss of core cooling. At Three Mile Island, that's exactly what happened.

Most reactor cores consist of fuel pellets encased in metallic tubes. The pellets are made of uranium dioxide, a ceramic with a high melting point. The metal tubes containing the pellets are referred to as fuel rods. These fuel rods are assembled into fuel bundles (also called fuel assemblies or fuel elements). An array of fuel assemblies comprises the reactor core. The reactor core is typically the size of a large bedroom and is where the heat-producing nuclear chain reaction takes place. The core is enclosed inside a pressure vessel with steel walls that are six to eight inches thick.

When operating normally, water flows through the fuel assemblies and carries the energy generated in the core to steam generators. On the day of the TMI accident, one of the steam generator feedwater pumps failed. This abnormal condition triggered an automatic shutdown of the reactor (just as it was supposed to). After termination of the nuclear chain reaction, a safety valve on the primary cooling system opened to relieve the pressure (also as it was supposed to), but the valve stuck open (which it wasn't supposed to) allowing cooling water to escape from the reactor vessel. Backup safety pumps were activated and flooded the core with cooling water (as they were supposed to), but control room operators, misled by faulty instrumentation readings, turned off these emergency core cooling pumps. Had the operators not turned them off, nobody today would associate the initials TMI with a nuclear accident. Without cooling, ultimately, about 45% of the core melted and of that, roughly a third slumped to the bottom of the reactor vessel. The containment building, a massive concrete and steel structure around the nuclear systems, performed as it was designed and prevented the release of radioactive contaminated water to the surroundings.

The accident was a devastating embarrassment for the nuclear industry and a financial disaster for the operating company, but almost a "non-event" in terms of public safety. Mind you, **I'm NOT saying there weren't tense days after the accident or that the public wasn't severely impacted by evacuations and even terrified in the early hours of the accident when information was scarce. Many in the public WERE rightfully frightened.** I'm only saying that the "defense-in-depth" safety philosophy of the nuclear industry and its regulators, a philosophy that's deeply impressed on every professional in the field, ensured that the accident didn't become a public health calamity. The safety systems and the defense-in-depth safety philosophy worked.

The Three Mile Island accident halted the expansion of nuclear power in the U.S. and began a process that's still ongoing today to design smaller, "fail-safe" or "passively-safe" nuclear plants that shut themselves down and require no active power sources to maintain adequate cooling of the core.

Chernobyl

The 1986 disaster at the **Chernobyl** nuclear plant in the Ukrainian SSR was vastly different from the 1979 accident at Three Mile Island. The Chernobyl accident was catastrophic. The death toll remains a matter of dispute. Most of the immediate deaths (more than thirty and probably less than one hundred) were a result of acute radiation exposure to workers at the plant who tried to extinguish the fire in the remnants of the burning core after the accident. Because the reactor had no containment structure surrounding the core, it's likely that thousands of individuals outside and downwind of the plant received radiation doses that took their lives in the months and years following their exposure or increased their risk of developing various forms of cancer.

The Soviet-designed reactor at Chernobyl could never have been licensed anywhere outside the former Soviet Union. The reactor was based on a design used in military reactors for producing materials to make nuclear weapons. It had a physically large reactor core that was inherently unstable. There were also characteristics of the core and control system that made the core susceptible to localized power transients especially at low power levels.

During an ill-conceived safety test and with the reactor at low power, the core underwent one of those localized power excursions. The rapid rise of pressure inside a limited number of cooling tubes resulted in a steam explosion sufficient to dislodge the upper structure of the reactor. When this occurred, the control rods became jammed and couldn't be inserted into the core. The cooling channels for the core were severed. In a matter of seconds, a second steam explosion and possibly a hydrogen gas explosion blew the core apart. About a quarter of the reactor fuel and graphite moderator were ejected from the reactor.

The Chernobyl plant had no containment structure around the core as is required in the West, so when the reactor blew itself apart, massive quantities of radioactive core material were released to the environment. Now exposed to the air, the core's graphite moderator caught fire. The plume from the fire carried radioactive material high into the atmosphere where the prevailing winds carried it to communities immediately downwind of the plant. These communities experienced the highest levels of radioactive contamination.

Nothing about the design of the Chernobyl plant would have been permitted in virtually any other country in the world: not the core design, not the control system, not the construction without a containment structure, not the operational procedures, etc. **The primary lesson from the Chernobyl accident was this: "Don't ever build or operate a nuclear reactor like this ever again, ever!"** Critics of nuclear power will always raise the specter of Chernobyl to scare the public even while fully understanding that the safety and design philosophy of Western nuclear power plants make such comparisons ludicrous.

Fukushima

The most recent nuclear disaster involved the six-unit **Fukushima Daiichi** nuclear station in Japan in 2011. A 9.0 earthquake forty-two miles offshore generated a forty-six-foot tsunami that arrived at the shores of the plant roughly fifty minutes after the earthquake. The three nuclear reactors at the site that were operating at the time of the earthquake were shut down safely moments after the earthquake. Offsite power (i.e., power coming into the plant from the outside) was unavailable because the regional power grid collapsed after the earthquake. The loss of offsite power is an occurrence that the plant was designed to withstand. The plant shifted successfully to onsite emergency power from diesel generators and began long-term cooling of the shutdown reactors.

A tsunami generated by an earthquake was considered in the plant design and a protective seawall was in place. However, the magnitude of the tsunami from the 9.0 earthquake was significantly greater than what had been designed for. The tsunami wave over-topped the seawall and disabled the emergency diesel generators. When that occurred, the plants were reduced to relying on battery power for instrumentation and limited equipment but didn't have enough power to run the cooling pumps. (Note: if the plant's emergency diesel generators and their fuel tanks had been situated on higher ground, the seaside community would still have been devastated by the tsunami, but the plants would have been spared and almost no one outside of Japan would even know the name Fukushima).

The reactors were shutdown successfully, but the reactor cores continued to generate decay heat from radioactive isotopes in the fuel. Portions of the cores in the three reactors that had been operating prior to the earthquake began to melt. Hydrogen gas, generated from chemical reactions inside the reactor/containment vessel was vented.

Eventually, the hydrogen gas accumulating inside the reactor building exploded – a **chemical** explosion, not a nuclear explosion.

In the absence of cooling, the reactor cores continued to heat up, portions melted and ultimately breached the pressure vessel, the internal containment structure, and the building shell. A small fraction of the radioactive core material escaped from the reactor vessels of the three cores, but nearly all of this material was still contained within the boundary of the reactor building.

The triggering event leading up to disabling the nuclear station was a **major** earthquake event that not only disabled the nuclear plants but caused wide-spread damage and destruction across Japan. **Unrelated** to the nuclear power plants and damage to the Fukushima reactors, more than **18,000 people** around Japan lost their lives in the earthquake and tsunami. In the final count, **one** nuclear plant worker lost his life due to radiation exposure received as a result of the accident. Residents near the plant were evacuated and an exclusion zone established. It's estimated that fifty people, mostly sick and elderly, may also have died as a result the stress they experienced during the evacuation (i.e., deaths **unrelated** to radiation exposure). Outside the Fukushima area, and again unrelated to the nuclear station, Japanese authorities estimate that another 1,000 people, also mostly sick and elderly, may have suffered premature deaths as a consequence of being displaced from their homes (for structural integrity issues, not nuclear-related) for an extended period of time.

Workers at the plant who came in contact with radioactive material after the accident were decontaminated. Aside from the one death attributed to radiation exposure, some workers and residents received low to moderate doses of radiation, which puts them at a minimally higher risk for some cancers. Japan will be monitoring these individuals for the remainder of their lives and will be engaged in clean-up activities at the plant for several decades.

The releases of radioactive material and the radiation doses received by workers and the public health consequences of both the Fukushima and TMI accidents were relatively minor. In Japan, the public health consequences of the nuclear accident at Fukushima were **many orders of magnitude** less than the consequences of the earthquake and tsunami.

High Level Waste Disposal

At the peak, there were just over 100 nuclear power plants operating in the U.S., supplying approximately 20% of the nation's electric power. Because nuclear power plants typically have the lowest operating cost of a utility's base load generation and because the characteristics of nuclear plants favor operation at a constant power level, most nuclear power plants are operated continuously at full power for a period of between twelve and eighteen months. After the fuel in the reactor reaches its design life, the plant is shut down for maintenance and refueling. At the end of the fuel cycle, roughly one-third of a reactor's fuel assemblies are removed, the remaining fuel assemblies are moved around inside the core and new fuel assemblies are inserted. The reactor vessel head is then bolted back in place and the core is ready to be restarted.

The fuel removed from the core is referred to as "spent fuel." It's highly radioactive immediately after removal from the core and is still generating a lot of heat. This decay heat diminishes quickly, but cooling is still required for an extended period of time. Freshly removed spent fuel is typically transferred to the spent-fuel pool adjacent to the reactor building. Natural circulation of water around the spent fuel assemblies is sufficient to keep the fuel bundles adequately cooled. Pumps and heat exchanges are typically only required to keep the water in the spent-fuel pool cool depending on how long the radioactive elements in the spent fuel have been decaying.

In addition to the spent fuel, there's a low-level radioactive waste stream from an operating nuclear power station. This waste stream consists of contaminated tools, protective clothing, gloves, and the like. This waste stream is analogous to the radioactive waste stream from a hospital, research facility or industrial enterprise that uses radioisotopes in various treatments or commercial activities. These low-level waste materials are typically disposed of in what might be described as sophisticated landfill sites. (My apologies to low level radioactive waste disposal site operators for this simplistic characterization of a complex and carefully monitored process).

Spent fuel is not low-level waste. The spent fuel contains isotopes of uranium and plutonium and their radioactive decay progeny and fission products, the "ashes" of the nuclear fission process. Collectively, this material includes elements that remain radioactive for thousands of years.

Originally, the plan for dealing with spent fuel was to mechanically and chemically reprocess the fuel assemblies to separate the useful components that remain in the fuel from materials that truly are waste products. The radioactive waste left over after extracting the useful and valuable materials from the spent fuel consists of materials that are radioactive, but collectively have a much shorter effective half-life and much smaller volume than unprocessed spent fuel. After reprocessing, this waste can be converted to a glass-like material, still highly radioactive, but physically stable and effectively as insoluble in water as the glass in a Coke bottle. This glass-like material was to be packaged in sophisticated containers and these waste cannisters buried deep underground in a dry, geologically stable location where it would remain out of contact with the biosphere while the radioactive isotopes decayed harmlessly away.

The significance of disposing of high-level waste in stable geologic formations far away from any groundwater is that such sites provide no pathway for the waste material to ever make its way back to the surface of the earth . . . forever. The fact that the material itself is insoluble in water means that even if water were to somehow find its way into the disposal repository, the radioactive materials would still not have a pathway back to the biosphere.

As the nuclear industry evolved (or failed to evolve in the U.S) "Plan B" for the spent fuel was to skip the reprocessing step and bury entire fuel elements in sophisticated containers, again in dry, geologically stable sites. Not reprocessing the spent fuel to remove the uranium and plutonium components, however, would result in a waste stream that would be much more voluminous and remain toxic for a much longer period of time than the waste from reprocessed spent fuel.

Fuel reprocessing is not mysterious, but it takes technical expertise. The U.S. reprocessed huge quantities of reactor fuel as part of the Manhattan Project. All of the nations that have nuclear weapons have reprocessed nuclear reactor fuel. North Korea is the most recent known addition to that club.

Regardless of how anyone views nuclear power as an energy resource, disposing of high-level nuclear waste is imperative. Even if every nuclear plant in the world were shut down today and not a single new nuclear power plant built anywhere in the world in the future, our generation would still have an **obligation** to safely dispose of the existing stockpile of high-level radioactive waste from civilian and military reactors.

From my perspective, **disposing of spent fuel without reprocessing it would be a travesty – an economic, a safety, a resource, and a moral travesty.** Unfortunately, no political constituency or U.S. political leader has emerged with the passion and commitment to fulfill this obligation that our generation has to future generations to reprocess all of the spent reactor fuel and safely dispose of the waste products. The nuclear industry in France, however, has been actively reprocessing the spent fuel from its reactors for years. One expedient option for the U.S. would be for the Department of Energy to contract with the French government to reprocess U.S. spent fuel. The contract should require the return to the U.S. of both the valuable heavy metal products (including uranium and plutonium) and the glass-like waste material. The heavy metals could be stockpiled for processing into fuel for future reactors **and** the transuranic and fission-product wastes processed for deep, geologic, permanent disposal in a U.S. facility. Regardless of how the task is accomplished (domestically or by international agreement), high-level nuclear waste disposal is a **moral responsibility** we have to future generations and disposing of **un-reprocessed** spent nuclear fuel should, in my opinion, never be considered as an option.

The Future of Nuclear Power

I seriously doubt that my discussion of the two primary objections to nuclear power, public safety and waste disposal, is going to change a lot of minds. However, for those who may have ruled out the nuclear option in the past, I hope these few pages may have provided a starting point for those willing to learn more about the technology and the role it could play in reducing or eliminating our reliance on fossil fuels. A viable nuclear power option will demand a heightened level of attentiveness by those given the responsibility to manage it. But if it's done correctly, the technology, with all its vices and virtues, could be a valuable tool in helping to limit the ravages of CGI.

Drum Roll, Please . . .

So, the question is: How does nuclear power, a carbon-free, proven, dependable electric generating technology factor into the battle to "beat" global warming?

Answer: **It doesn't.**

Nuclear power is not merely a good carbon-free power option for the future, it's the **only option** that's widely deployable, non-intermittent, carbon-free, and has a virtually inexhaustible supply of fuel. Nevertheless, the public isn't ready to usher in the "second nuclear coming" and, from what I can tell, won't be ready until literally every other energy option has been tried. Unfortunately, when the public finally

> *The public isn't ready to usher in the "second nuclear coming" and, from what I can tell, won't be ready until every other energy option has been tried and found wanting. Nuclear power in the U.S. will remain the energy option of last resort.*

does come around to supporting the deployment of new, advanced nuclear plants, it will be too late. **The time required to build and demonstrate the new, advanced reactor designs, learn from these demonstration projects, refine the designs and then build the follow-on power plants in numbers sufficient to make a significant dent in our CO_2 emissions will be just too long.** The best we can hope for in regard to nuclear power is that the **currently operating nuclear plants in the U.S. and around the world won't be shut down prematurely and that they will be granted license extensions, if technical criteria are met, when they reach the end of their current operating licenses.**

Until new nuclear plants are built the number of operating nuclear reactors in the U.S. will slowly decrease, even with license extensions. As these carbon-free electric generating resources are shut down, alternative generating capacity will have to be brought online. Solar and wind power facilities may be deployed to offset the loss of the nuclear generating capacity, but there will be no reduction in CO_2 emissions with this shift because the energy generated at the nuclear power plants was already carbon-free.

Nuclear power will play an expanded role in the world's energy mix and efforts to "beat" global warming, but not until 2050 or 2075. That schedule could be accelerated if we make a national commitment to build two demonstration nuclear power plants (based on two different technologies) in this decade. Almost as important as the new reactor technology, however, we also need to demonstrate that we can build and license these two demonstration reactors quickly and at a reasonable cost. Before a single shovel hits the ground, teams of cost engineers need to scour the plans looking for ways to simplify the designs and reduce construction costs. Plant designers and regulators in the 1970s and '80s piled on requirements that lengthened construction times and raised costs. They acted as if the initial cost of the plant didn't matter because the fuel was so cheap. They were wrong. Many of the newer nuclear concepts incorporate passive safety features. Designers need to capitalize on those characteristics to develop designs that are less costly, and regulators need to credit those features in their oversight responsibility.

We used to be able to build things quickly in this country. The Empire State Building, for example, the tallest building of its era, was built in 410 days. The beams from the steel mills in Pittsburgh where they were manufactured were reportedly still warm when they arrived at the construction site in Manhattan. The two new nuclear demonstration plants that eventually get built as part of an aggressive response to global warming need to emerge from a substantially revamped licensing and construction process.

Because the public is so wary and because outreach efforts by the nuclear industry have been so inadequate, nuclear power won't make a comeback until massive quantities of solar and wind power generation and energy storage capacity have been deployed and found to be inadequate, severely compromise the reliability of the grid and/or are too expensive.

Then, and only then, as attempts to drive down CO_2 emissions stall and the shortcomings of a solar-centric grid become apparent will the public finally understand that "going green" also means "going nuclear." Several prominent former "anti-nuke" activists have come around to recognize the role that nuclear power must play in reducing CO_2 emissions, but regrettably, most of the world (except China) will continue to treat nuclear power as an undesirable last resort. As long as that perception prevails, nuclear power will not play a significant role in combating global warming. In spite of the Fukushima accident, Japan, for example, will eventually return to the nuclear option out of necessity and are likely to do so long before the U.S. Once nuclear power is again at least accepted by the public, it will take another one or two decades before the number of operating nuclear power plants worldwide is sufficient to take a meaningful bite out of CO_2 emissions. By then, the environmental impact of global warming will already be severe. Deploying nuclear power plants in the latter half of this century will be an attempt to lessen the further deterioration of our global living space, not "beat" global warming. That battle will have already been lost.

Action Items

Have we "snatched defeat from the jaws of victory" by effectively turning our backs on nuclear power? Will nuclear power be the technology future generations rely on? Will future generations look back at this period and wonder "what were they thinking?" Here are two related project initiatives that could, in the distant future, help address global warming stresses.

1. Build two demonstration nuclear power plants (based on two different technologies) within seven years.

2. Begin reprocessing spent nuclear fuel (domestically or through an international agreement) within five years. Make the permanent nuclear waste disposal repository operational within six years. (Note: These nuclear-waste-related action items won't displace any CO_2 emissions, but they may neutralize the nuclear waste argument that's been preventing a number of people from advocating an expanded nuclear dependence).

8 RENEWABLES: FREE ENERGY FOR ALL

"The answer, my friend, is blowing in the wind."
Song Lyric, Bob Dylan, 1962

"Here comes the sun."
Song Lyric, The Beatles, 1969

The "Other" Renewables

Wind turbines and solar panels are the two technologies that most people think of as renewable energy technologies. But there are others, and it may be useful to review these alternatives (and dispense with them) before narrowing this discussion to the two most prolific of the renewable technologies.

Biomass power generation (i.e., garbage incineration, not ethanol) is an option that will no doubt play a larger role as part of a future energy mix. It has the advantage of reducing the amount of material going to landfills and, therefore, ultimately reducing the amount of methane those disposal sites emit. Burning biomass does, of course, produce CO_2, but this is carbon that had previously been extracted from the atmosphere by plant life and is being returned to the atmosphere when the biomass is burned. Biomass burners, however, can be complicated and expensive because they generally require sophisticated pollution abatement equipment.

The fuel for these garbage-burning plants is basically anything we discard. However, that includes a wide range of materials that could become toxic or carcinogenic when incinerated if they are not separated out of the feed stream or scrubbed out of the exhaust. These plants also face siting restrictions and are the object of NIMBY concerns, i.e., "not in my backyard." Because of all the added steps and safeguards (i.e., collecting the "fuel"/garbage, waste stream separation and fuel preparation, combustion, exhaust cleanup, and waste disposal), when this power generation option is utilized, the cost of power from it will be several times higher than energy from a natural gas burning power plant. For all these reasons, this option will probably be used, but not as much as it could or should be.

Another renewable resource is geothermal power generation. This technology involves extracting steam or hot water from deep in the earth, bringing that energy-bearing medium to the surface, processing it, transferring the energy from the original hot "medium" to a secondary working fluid (usually steam) to drive a turbine or for use in district heating applications. Where rich geothermal resources exist, for example in Iceland, this option is viable, but favorable geothermal sites are comparatively limited worldwide. In addition, it's a challenging resource. Geothermal liquids contain corrosive and often noxious chemicals such as hydrogen sulfide (think rotten eggs). So, while the energy resource is free, even in geothermally rich areas the capital investment, maintenance expenses and environmental requirements all drive up the cost of energy from this resource compared to others. The technology is being used, and will be used even more in the future, but only in limited areas.

Wave power generators and tidal power plants have also been the object of a lot of research and development for decades, but there's a reason these resources haven't been widely deployed. As with wind and solar, wave power generators use an energy resource that's free, but has such a low energy density that the capital investment per kW of generation capacity is high. In addition, anyone who's ever owned a boat in seawater knows how unforgiving the ocean environment can be. Finally, protecting ocean-based power generation from damage during storm events adds to the cost and risk associated with these facilities. Electrical energy based on extracting energy from ocean waves will never contribute significantly to the world's generating resource mix. If it could, it would already be doing so.

There is another form of ocean power that takes advantage of especially high tides in unique geophysical locations, but because these locations are so unique, this method of extracting energy from the movement of ocean water will also not contribute significantly to global energy resources.

As energy becomes more expensive, lots of resource options that might **not** have made economic sense previously will gradually come back on the table. Micro-hydro power generation is one example. This form of power generation exploits smaller flows of water. The technology is proven and readily available. It's "just" a matter of economics. That being said, because these resources tend to be small (otherwise they would have already been developed), this too is a technology that will probably be utilized to a greater extent in the future (as energy costs rise) but will make only a relatively minor contribution to the portfolio of future carbon-free power resources.

As noted above, changes in the price of energy alter the viability equation for a range of supply side **and** demand side energy options. Declining oil prices (yes, they will decline, but they will also rise again and decline again) cause producers to shut down higher-cost operations such as oil extraction from tar sands and oil shale. Increases in energy prices favor investments in conservation technologies and other behavioral changes. All lifestyle changes and hardware installations that conserve energy (e.g., walking to the store instead of driving, leaving the air conditioner off, installing double or triple pane windows, adding building insulation, etc.) will be doubly beneficial: The energy that's not "consumed" for whatever reason through conservation, will be available for some other use.

Homeowners should anticipate that they will be paying dramatically more for energy in the future. Any home improvement decisions they make today should anticipate those higher energy prices in the future. Such a cost-benefit analysis should probably assume that energy in the future will cost

> **Homeowners should anticipate that they will be paying dramatically more for energy in the future.**

at least three times what it does today. Similarly, building and appliance energy efficiency standards should anticipate much higher future energy costs, and those standards should be made correspondingly more stringent.

From this point forward, wave power, biomass, micro-hydro, geothermal, garbage-burners, and all other

> **Wind and solar will edge out all the other renewable resources in the market.**

renewable resources will have to compete with wind and solar generation. These are the two technologies that are proven and are more economical. This is a case where one or two of the renewable resources are likely to edge out all the others in a competitive market (unless those other carbon-free alternatives offer some "value-adding" feature, such as not being intermittent). That statement bears repeating: **the lowest cost renewables will edge out all the other renewables in the marketplace.**

Solar and Wind Power Take Center Stage

Solar panels and wind turbines are the most universally deployable and cost-effective of the renewable energy technologies and are, therefore, the most prolific, the fastest growing and the most visible. Solar panels have the added advantage of being readily deployable in small increments close to the place where the energy is being used (assuming there's an adequate number of sunny days to justify the investment).

The vast majority of solar energy in the U.S. is being generated below 37° north latitude, i.e., in the southern half of the country – south of San Francisco on the west coast, Missouri in the middle part of the country and Virginia on the east coast. Exhibit 8.1 illustrates the U.S. average annual "solar resource" available to be captured for electric energy generation.

This map clearly demonstrates a number of things that are particularly important if solar power is to play a significant role in freeing us from fossil fuels. First, the most efficient solar installations will be located in the Southwest. This is also the region of the country that experiences the highest temperatures. The hotter it gets, the greater the fraction of solar energy produced in the Southwest that will be consumed locally to run air conditioners and power swimming pool pumps. Less will be available to export to other parts of the country except during the cooler months. For the energy that's exported, this map also suggests why there will have to be a massive investment in the high voltage electric transmission system (to transport energy to the north and to the east) to fully exploit this resource.

A 1 kW set of solar panels operating at maximum capacity for one hour produces 1 kWh. A single "residential" solar panel is usually around 0.33 kW. A typical residential solar installation may consist of 20 to 40 panels and have a peak power production capability of between 7 kW and 15 kW. A 1000 MWe thermal power plant (1,000,000 kW) powered by coal, natural gas or nuclear, (i.e., a large power plant), operating at full power for one hour will produce 1,000,000 kWh. Operating at full power, the plant would produce 24,000,000 kWh in a day. A residential solar panel installation consisting of thirty solar panels would typically produce about 10 kW

Exhibit 8.1
Solar Resource Intensity*

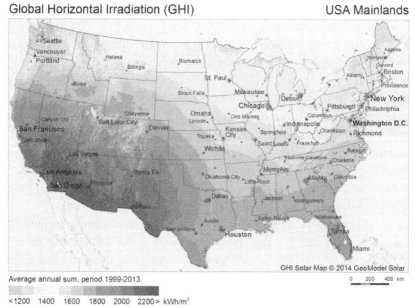

Global Horizontal Irradiation (GHI)   USA Mainlands

Average annual sum, period 1999-2013

<1200 1400 1600 1800 2000 2200> kWh/m²

GHI Solar Map © 2014 GeoModel Solar

0   200   400 km

(*Graphic used by attribution:
GHI – Total Solar Radiation incident on a horizontal surface.
SolarGIS © 2014 GeoModel Solar, CC BY-SA 3.0,
https://commons.wikimedia.org/w/index.php?curid=36758775)

of peak power. Averaged over the year, such a system would typically produce an average of between 40 and 50 kWh per day.

A 1000 MWe thermal power plant will produce 0 kWh when it's down for maintenance or disconnected from the grid for any reason. A 10 kW solar system will produce 0 kWh after sundown and before sunrise.

A typical home refrigerator draws between 0.15 and 0.3 kW. If the refrigerator compressor and fans run for a total of 8 hours in a 24-hour period, the refrigerator will consume between 1.2 and 2.4 kWh per day or roughly between 450 and 850 kWh per year. A medium-size residential central air conditioning system (3 tons) may use roughly 3.25 kW when it's running. If the A/C system runs 9 hours a day during the hot part of the year, the system will consume approximately

30 kWh per day. If air conditioning is needed four months a year, the A/C system will consume approximately 3,500 kWh per year.

The energy usage for these two energy-consuming functions, the refrigerator and the air conditioning system, are both measured in kilowatt hours. As previously discussed however, from the utility perspective, these two loads couldn't be more different. The load for the sum of all refrigerators on the system turning on and off at random all day long, is comparatively constant 24 hours a day, 365 days a year. The most intense air conditioning load, on the other hand, is concentrated in a limited number of hours (six to nine hours), in a limited time of the day (10 am until six pm), for a limited number of months of the year (four to six months) for a limited period of the year (mostly summer). The energy generating assets a utility would put in place to serve these two different kinds of loads most efficiently would be very different.

The daily power production profile for thousands of residential solar systems having a combined nominal maximum rating of 1000 MWe (perhaps 100,000 homes) on a sunny day in summer is shown in Exhibit 8.2.

Exhibit 8.2

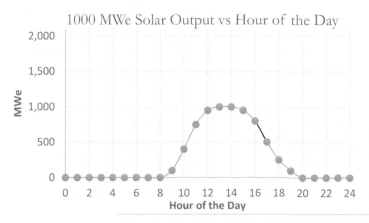

Production begins at a low level just after sunrise and builds as the sun climbs higher in the sky, peaks in the early afternoon and declines as the sun sinks lower in the sky in the afternoon.

Most residential solar installations are not as efficient as shown in Exhibit 8.2. Residential installations on existing roof lines are seldom optimally positioned to maximize solar energy production. In addition, residential installations are typically fixed in place. They are "optimal" (i.e., as good as they can be) on only one day of the year. Every other day of the year, they are slightly less efficient because the angle at which the sun's rays strike the solar panels changes day-by-day. Some homeowners may optimize their installations for maximum solar generation during the winter months while others may choose to optimize for maximum energy production in the summer. In most cases, the orientation of the roof makes the decision for the homeowner. Most homeowners probably don't have a complete understanding of what their contractor installed on their roof. I say this from the experience of walking the neighborhood and spotting solar panels positioned on steep north and northeast facing roofs!

Adding significant quantities of homeowner-owned solar generating capacity interconnected to the grid has a big impact on utility operations. Fast forward more than 30 years from the time when the power demand on the Southern California Edison system looked like Exhibit 6.2, to today where a simulated customer demand profile from the grid on a typical hot summer day might look like Exhibit 8.3. In this case, "typical" means the sun is shining, the skies are clear, and a gentle wind is blowing. In other words, a Chamber-of-Commerce kind of a day.

The shape and amplitude of this power demand profile are quite different from the summer curve shown in Exhibit 6.2. Due to population growth, increased economic activity and a higher intensity of energy usage by most customers, the magnitude of total power usage by all consumers is much higher today, but more importantly, the shape of the curve, especially in the middle part of the day, is quite different. This difference reflects the surge in the installation of residential solar power generation.

The "self-generated" solar electric energy is not extracted from the grid, but it is, nevertheless, energy the customers are either using for their own needs or delivering to the grid for credit from the utility. Adding together the customer-generated energy and the energy customers receive from the grid, the **total** hourly energy usage has the same general shape it did 30 years ago. This is shown in Exhibit 8.4 where the self-generated energy component of the total supply picture is shown in the cross-hatched area. On a sunny day, the power residential customers may pull from the grid might look like the lower line in Exhibit 8.4 while their demand from the grid on a cloudy day might look more like the upper line.

Exhibit 8.3
2020 Simulated Grid Demand in the Era of Renewables

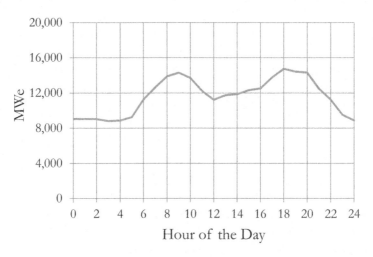

The simulated night-time power demand in Exhibit 8.4 is about 50% of the maximum daytime total power (self-generated plus grid-supplied) and roughly 65% of the maximum demand on the utility's grid. (These percentages will vary by region). This self-generated energy has enormous implications for the operation of the grid and the transition to a decarbonized future.

First, when customer-owned generating resources (i.e., rooftop solar installations) are connected to the grid in massive numbers, the energy generated by and used by the customer is **energy the utility company doesn't sell.** If the utility company (or the owner of the electric generating station) doesn't sell as many units of energy (i.e., kilowatt-hours), all of the utility's fixed expenses (i.e., "overhead" charges and charges to recover the costs of previously constructed power generating plants that are now no longer being used as heavily as originally planned) are allocated across fewer kilowatt- hours sold. This drives up the per kilowatt-hour price paid by all customers.

A second impact of large quantities of self-generation (rooftop solar) on utility operations occurs when customers who normally produce some or all of their own power are forced to pull all of their power from the grid. Such a situation would arise on partly cloudy or overcast days. On those days, several thousand megawatts of solar generating capacity will be unavailable. Utility generating stations

Exhibit 8.4
2020 Simulated Grid Demand
(With and Without Customer Self-Generation)

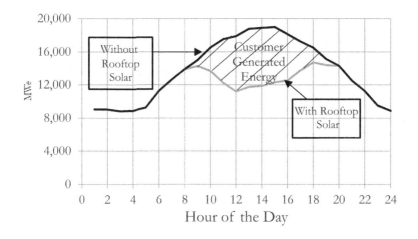

(most likely powered by fossil fuels), that had effectively been sitting idle on sunny days will now be called upon to serve the demand. At this point, the law of supply and demand kicks in, just the way it did during The Great Texas Freeze in 2021. Substantially reduced energy supplies will give rise to higher cost energy from these backup generators. When this occurs, the cost of energy for all customers on the grid goes up. To offset the cost of acquiring and maintaining back up power supplies for these intermittent customers who only pull this much energy from the grid when clouds roll in, utilities may rightly demand a hefty grid connection premium.

One of the secondary effects of residential solar installations is that many customers are using more energy for air conditioning. Prior to making the investment in solar panels, many customers chose to forgo the comforts of air conditioning at the hottest times of the day to save money. After installing solar panels, there is anecdotal evidence that many of them choose to operate their air conditioning systems more frequently than they had before installing the solar panels. Who would blame them? They feel entitled to use their air conditioners more because they have solar panels and/or because the cost of running their air conditioning "feels" like it's less. This behavioral shift has increased the total number of kWhs (self-generated plus grid-generated) consumed to provide air conditioning to increase at a rate greater than population growth. This, in effect, represents a "new demand" that didn't exist prior to the installation of the solar panels. On especially hot days this behavior makes less of the capacity the solar panel owners installed on their rooftops available to displace other energy uses in their homes or fossil-fuel energy generated by the utility.

One final thought about air conditioning. The number of days with temperatures exceeding 90° F has been increasing all around the country. As the planet warms, one local effect is that air conditioning electrical loads are likely to increase in both magnitude and duration (i.e., the air conditioning season will start earlier in the spring, extend longer into the fall and require more hours of cooling on hot days). In other words, decarbonizing the grid is a moving target that's becoming more difficult as the planet warms.

Wind Turbines

Unlike solar panels, most large, modern wind turbines are sited in remote areas or, more recently, offshore. They are also typically sited farther away from major load centers because of environmental issues and the fact that the locations with the best wind resource (i.e., strong consistent winds), aren't typically where most people want to live.

Not long ago, a large wind turbine may have been rated at 1 MWe. Today, the most efficient and most cost-effective wind generators are larger, between 2 to 10 MWe. (Machines as large as 15 MWe are being tested, but there is little or no operational or durability data for them). The average wind turbine in the U.S. is just over 2 MWe. (This value is lower than the newest ones being added to the grid today because most of the wind turbines in operation across the U.S. today were erected before the latest models were developed). Unlike solar generation that has a partially predictable intermittency (i.e., nighttime), wind power generators can operate at any time of the day or night when the wind is blowing above a certain minimum speed and no higher than the upper design limit for the unit. This gives wind a significant "leg-up" over its partner renewable, solar power, but it still carries the requirement for massive investments in storage capacity and transmission lines.

Solar Power – A Deeper Dive

Many residential and commercial solar installations are subsidized by the government or by other utility customers through installation rebates, tax credits and/or favorable buy-back rates mandated by public utility commissions. Some of this beneficial treatment was designed to promote the deployment of solar capacity. Here in Southern California, it worked. In general, though, subsidies of any kind distort free markets and, unless administered carefully, can send erroneous pricing signals to consumers.

Well-designed subsidies encourage behaviors that benefit society. But of course, what one individual regards as beneficial might be abhorrent to someone else. For example, low emissions vehicles with only one occupant are allowed access to the HOV ("carpool") lanes in California. This benefit, i.e., subsidy, clearly motivated many commuters to purchase clean vehicles. I might believe that allowing single-occupant, zero tailpipe-emissions vehicles to use the carpool lane is a beneficial outcome for society, while a car pooler in a gasoline-powered automobile with four passengers might be frustrated that so many single-occupant, clean air vehicles are clogging up the carpool lanes.

The subsidies and tax breaks for solar panels helped the nascent solar industry grow sufficiently to achieve the cost reductions associated with mass production. The success of the industry in California suggests that the time for phasing out those subsidies may have been reached. Solar panels are now cost competitive with fossil fuels on a "kWh-produced" basis. (I'll explain what I mean by this qualification shortly.) With the proper energy buy back rates from utilities, solar panels may now be economically deployable nearly anywhere in the U.S. without upfront subsidies.

It's difficult to compare the cost of energy from solar systems to that of energy from an electric utility on an apples-to-apples basis. Solar panel systems come with some hard-to-quantify and often overlooked costs.

Solar panels need to be maintained. (Most residential solar system owners don't, but they could get better performance from their investment if they did). At the very least, the surfaces of the panels should be cleaned periodically to achieve peak performance. The electrical equipment to convert DC power to AC power and the interface with the local utility, may need periodic testing, maintenance and/or replacement. Roof maintenance may require removal of the solar components but reinstalling the original panels after a roof repair may not make economic sense. Efficiency improvements and lifetime limitations of the original panels may make it advantageous for homeowners to purchase new equipment. All these activities, while infrequent, incur costs the homeowner should consider when comparing the cost of owning their own solar generation system vs. purchasing power from the utility.

Stormy Weather

Another issue relating to solar and wind generation is their vulnerability to high winds and other severe weather. In 2019, monster hurricane Dorian churned up the U.S. east coast from Florida to North Carolina, with peak winds approaching 200 mph. Fortunately, the main force of the storm tracked offshore, but it raises the question: what if Dorian had penetrated 100 miles deeper into Florida and Georgia before turning north and tracking up the Atlantic coast? Aside from the widespread property damage unrelated to the regional electric grid, the storm could have destroyed or damaged virtually every rooftop solar installation in its path. The solar panels themselves could have become flying debris that inflicted even further damage to other structures.

Wind is only one element of severe weather. Snow and ice storms will also decrease the capacity factor of wind and solar facilities. These are real-world issues that must be considered when determining regions of the country (and regions of the globe) where different types of renewable resources can or should be sited. With so many wind and solar installations being deployed, we're rapidly gaining additional experience on the durability of these facilities in severe weather environments.

[*Input] – If any entity has documented the survivability of solar and wind resources after extreme weather events, please post it to the website, BlueOasisNoMore.com [*Input – End].

Aside from the durability and survivability of solar and wind installations, **the big problem with both solar and wind power is that they are intermittent energy resources**. Solar panels produce no energy at night, of course, but they're also less productive when it's overcast, when the panels are shaded or dirty, and when the angle of the solar panels is less than optimal.

> *The intermittent nature of the resource is the "Achilles heel" of both wind and solar resources.*

There is one version of solar power designed to compensate for the daily unavailability of the resource at night. These thermal solar power generation facilities concentrate and collect the sun's energy by heating a working fluid that's stored in large tanks. Energy can be extracted from the hot working fluid at night to produce steam which can be used to generate electricity, but not for extended periods and not without the added expense, thermal inefficiencies and design complication of a thermal storage reservoir and thermal energy conversion equipment.

The U.S. Energy Information Agency (eia.gov) collects data on the performance of all energy resources. Capacity factor data for 2019 for the primary renewable technologies were: hydroelectric – 41%, wind power – 34% and utility-scale solar photovoltaic – 24%. The online environmental magazine *Yale Environment 360* (e360.yale.edu) reported similar 2016 performance characteristics: hydroelectric generation -- 38.2%, wind turbines --34.5% and utility-scale solar arrays -- 25.1%. These numbers may vary in any given year but will probably not change significantly in the near term. In the long term, however, all three could drop for different reasons. Decreased precipitation in the major drainage basins in the West will decrease the amount of water available to flow through hydroelectric generators and the capacity factors for wind and solar power systems are likely to decrease because future deployments may be in less favorable locales. For example, a solar system in Seattle would generate only half as much energy annually as that same system in Phoenix

If a generating resource has a lower capacity factor, it means that more facilities of that type must be built to generate a given amount of energy over a fixed period of time. In the case of wind turbines, for example, with an average capacity factor of 34%, three wind turbines would have to be erected to generate as much energy as just one of those turbines spinning continuously.

In addition to the low capacity factor of the wind turbines, an even more important consideration is that the utility has no control over when those turbines are generating energy. At times, all three wind turbines might be spinning while at others, all three may be idle.

These same considerations also apply to solar panels, except that solar systems also have the **predictable unavailability** we call "night." The critical link connecting intermittent power generating resources to the instantaneous demand on a modern utility grid is energy storage, the topic addressed in the next chapter.

However, before leaving the issue of capacity factor, it's worth noting that the capacity factor for utility-scale solar arrays is much higher than your typical residential rooftop system. The larger, utility-scale installations are well maintained, and their panels are optimally positioned to maximize energy generation. The panels of many utility-scale arrays are even mounted on motorized platforms that keep the solar panels aimed directly at the sun throughout the day every day of the year. By contrast, residential rooftop arrays are seldom well maintained and are constrained by the direction and angle of the roof. The net effect is that residential rooftop solar arrays typically have capacity factors hovering around 16%, i.e., one third lower than optimally positioned utility-scale solar arrays.

Transmission Lines to the Rescue?

Some of the intermittency of wind and solar resources can be diminished by relying on high voltage transmission lines to move power around large geographic areas. This is done today whenever excess power or more economical power is available in one area and needed in another, often hundreds of miles away. Utilities and/or electric system operators use weather and geophysical data to model the likely availability of intermittent regional power supplies and needs. In the case of solar capacity, for example, excess solar energy in one area with clear skies may be available for transmission to another area that's normally adequately supplied by local solar resources but is currently overcast and in need of additional supplies.

Before leaving this brief mention of transmission lines, I need to emphasize just how important these lines will be in terms of unlocking the potential of both wind and solar power generation. They will be the key to minimizing the quantity of bulk energy storage capacity that will need to be constructed to maintain grid reliability north of 99% and move energy from wind-resource-rich areas to population centers. Many people consider these large transmission structures a blight on the landscape. As the era of renewables unfolds, these lines will come to be recognized as "beautifully-ugly," the way some people might see hydroelectric dams. As hard as utilities may try, and as skilled as they may be at using the transmission line network to even out the intermittent nature of solar and wind generation, these resources are still intermittent and there will be times when meteorological conditions will render the intermittent generation resources inadequate to serve the electric demand.

Intermittent Energy: What's It Worth?

I previously noted that solar panels are cost competitive with fossil fuels, but only on a "kWh-produced" basis. I need to explain what I meant by that.

What I was alluding to is the difference between energy from a base-loaded generating unit such as a natural-gas-fired power plant and an intermittent power source such as a solar installation. Each may produce a kWh of energy at a cost ten cents during the daytime. Viewed this way, the two resources may appear to be cost competitive, but that's a false comparison.

A base-loaded, natural-gas-burning generating station is capable of delivering energy to the grid 24/7; i.e., it has a high availability factor. A solar array does not. In order to compare the cost of the energy coming from the two generating resources in a meaningful way, it's necessary to add the cost of a bulk energy storage facility and the cost of the added renewable generating capacity to charge that storage medium, to the base cost energy from the solar system. Doing so captures the actual cost of solar energy capable of providing energy 24/7 in the same way a natural-gas-burning generating station can. When compared on this basis, the cost of energy from the solar/storage installation system will go up dramatically.

By way of analogy, consider the person who chooses to forego car ownership because they can ride a bicycle to work for a dime a day. Riding to work is an economical and healthy commuting choice . . . on sunny days. But what happens when it rains, snows, or reaches 110° F and the bike rider has to take an Uber? Claiming their commuting costs are a dime a day ignores the expenses incurred when getting around on two wheels isn't an option. This same principle applies to solar power. Claiming solar power is cost competitive with energy supply sources that are not intermittent is a willful choice to ignore the additional costs solar power carries with it to serve real world customer demand.

Electric Rate Design in the Era of Intermittent Renewables

Large quantities of customer-owned solar panels reduce the number of kWhs sold by the utility. None of the utility's fixed costs (administration, maintenance, personnel, the transmission and distribution infrastructure, company vehicles, profit, etc.) have similarly decreased. Any business, not just utilities, that sells fewer units of whatever it's selling, is forced to increase prices on the products that it does sell to cover these fixed costs.

This is already happening in my home state of California and probably everywhere residential solar installations have boomed. One of the contributors to increasing electricity rates is these higher fixed charges due to lower sales. But in the case of solar generation, there's the added impact of the credits the utility companies are "paying" to residential solar producers for the excess energy they deliver to the grid. If the credits to solar-system owners are higher than what the utility would have to pay other suppliers for the same energy at the same time, then those added expenses also have to be passed along to all customers. These two sources of rate increases fall into the general category of "cross-subsidization."

Cross subsidization is the practice whereby one class of utility customers, for example, industrial customers, pay some of the costs incurred by the utility to deliver energy to another class of customers, for example, residential customers. In the interest of fairness, utilities and regulatory commissions try to minimize cross subsidization by designing rate structures that allocate to each class of customers (to the extent possible) the costs associated with delivering energy to that customer class.

Intentionally or unintentionally, public utility commissions and the utilities have, I believe, developed rate structures that cause one set of residential customers (residential customers **without** solar panels) to subsidize another class of customers (those **with** solar panels). Some consumer groups advocating on behalf of low income ratepayers (who typically don't own solar panels) have filed complaints with the California PUC to address the issue of intra-customer-class cross-subsidization inequities. The CPUC has conducted hearings to solicit input on the issue. The utilities want to lower the payments to residential customers who own solar systems while solar system installation companies and solar systems owners are crying "foul."

The easiest and most equitable way to fairly compensate solar panel owners for the energy they deliver to the grid would be for the system operator to pay them the same rate they pay to all other energy providers at the same time of the day. Modern electronic metering makes this possible. This would eliminate any intentional or unintentional intra-class cross-subsidization in rate design. The net result would be a more equitable pricing approach for all customers. As more solar capacity is deployed in a region, it might also mean that on all but the hottest days, solar system owners are not paid as well in the future as they have been up until now. In short, they could be the victims of their own success. This issue will become an important (and contentious) consideration in the future. Not only would solar system owners receive less for the excess energy they generate and deliver to the grid on non-peak energy demand days, the value of the energy produced by wind turbines would be similarly devalued at times when excess intermittent energy of all kinds is available.

During the transition to a carbon-free grid, the utility will have to maintain its primary energy delivery system, while simultaneously supporting a "shadow" system (through contracts and rate design) of renewables and storage. This is somewhat similar to the homeowner who buys a new home before selling their current home. During the transition, the homeowner has expenses associated with their new home, e.g., mortgage payments, utilities, property taxes, etc. while continuing to have the same kinds of expenses for the house that's on the market. In the transition to a carbon-free utility grid, the utility will have all of the expenses (either directly or indirectly) of its fossil-fuel generating facilities plus the expenses associated with the carbon-free generating assets that feed into its system.

Action Items

The transition to a carbon-free grid heavily dependent upon intermittent generating resources requires a different approach to utility operations and electric rate design.

1. Credit residential solar power producers at the same rate paid to other power producers.

2. Implement dynamic energy pricing so that customers know the cost of the energy they're consuming at the time they consume it.

9 BULK ENERGY STORAGE

"Ask not [how many solar panels we've installed],
Ask [how much storage capacity we've built]."

With apologies to President John F. Kennedy,
a rewriting of the most-quoted line
from his inaugural address, 1961

Fossil fuels (coal, oil, and natural gas) are "energy dense" and can be stored – not characteristics of solar and wind power. As discussed in the prior chapter, these renewables are intermittent, diffuse, and cannot be stored without first converting them to some other form. It would be nice if we could deliver a gallon of sunlight or a cubic foot of wind, but we can't. These differences (energy density and storability) between carbon-based and carbon-free energy sources have enormous implications for the challenge of decarbonizing the world energy economy.

In order to "beat" global warming, we'll have to electrify virtually every human activity with carbon-free electric generating technologies. But, with nuclear power "off the table," we (effectively) only have intermittent energy resources to accomplish this task and that leaves us only a few pathways forward. We could:

1. Abandon the requirement for extremely high reliability of our electrical grid, i.e., accept regular blackouts as the new normal, or
2. Operate fossil-fuel-burning power plants on an as-needed basis, or
3. Build **a lot of electrical energy storage capacity and an even greater amount of renewable generating capacity** solely dedicated to charging that storage, or
4. A combination of all three.

From the perspective of global warming, Option #3 is the most desirable; Option #2 is the worst. Option #1 has economic and lifestyle consequences that would be intolerable. Option #4 (surprise, surprise) is how this will play out.

Option #1 - Sacrifice Reliability

It's worth noting that high quality, reliable electric service has been available in the developed world for only a little over 120 years. Even though this is a fairly recent "luxury," it's unlikely that the American economy, or the economies of other developed countries, would tolerate an electrical grid with a significantly diminished reliability. Can you imagine an America where power shuts off frequently and unpredictably for hours at a time (as happens daily in many developing countries)? I don't think so.

That said, attempting to maintain the same high level of reliability of electric service that we currently enjoy in the future when fossil fuels are not the primary fuel for the generation of electricity **may not always** make economic sense. Two examples, one from California and one from Texas:

In 2000-2001, several large energy companies, Enron among them, manipulated the electric energy supply market in California and dramatically drove up the spot price of power by as much as a factor of 20 for short periods. Conditioned to "never let the lights go out," electric system operators agreed to pay a high price for the last increment of power needed to keep all customers fully supplied. In so doing they committed the utilities to pay all wholesale power producers delivering power to the grid at that moment the same high price being paid for that last increment of supply. The consequence of accepting that last increment of high-priced power was a dramatic spike in energy costs for all customers. Because retail electricity charges were capped, the utilities were stuck with the losses. The state's largest utility, PG&E, was driven into bankruptcy and the second largest, SCE, came close to the same fate.

In this case, the system operators made an operational choice, based on their past experience and training, to maintain system reliability at all costs. They could have rejected the high-priced power and cut demand using every tool at their disposal including localized blackouts or brown outs. Doing so would have been highly controversial but may have been economically justified.

More recently, the February 2021 deep freeze in Texas resulted in sky-high electric bills for many Texans. As far as I know, none of the suppliers in the Texas spot market for electricity manipulated supplies to boost profits. Mother Nature took care of that for them. The energy producers simply took advantage of the most basic rule of economics. As electric supplies shrank and power demand soared, they charged (literally) as much as they wanted for their product. In some cases, the wholesale price for electricity went as high as 180 times pre-freeze levels. As miserable as conditions were, many customers would have accepted being blacked out to avoid the dramatically higher electric bills had they been given the choice.

This "quality vs. price" choice is one that most of us deal with regularly. When shopping for groceries, I may want to enjoy prime filet mignon beef every night, but I don't. I might buy chicken or tofu instead, until my more-expensive first choice goes on sale. I make a conscious decision to forgo the quality and enjoyment of the filet in favor of the lower-cost alternative.

Similarly, accepting a minor reduction in electric service reliability may be an option that system operators may need to consider **under certain circumstances.** There are expenses associated with electrical outages, but there are also expenses associated with building resources and infrastructure to meet exceedingly high reliability requirements. It's a trade-off. The current reliability standards were developed when backup energy resources were much less expensive than they will be in the "green" era. In the future we may have to learn how to accommodate a slightly less reliable grid.

Nevertheless, Americans would riot in the streets if their air conditioners didn't roar to life when they flipped the switch.

> *Americans would riot in the streets if their air conditioners didn't roar to life when they flipped the switch.*

Option #2 – Continue Using Fossil Fuels

The second option for maintaining system reliability after a significant quantity of intermittent resources are deployed on the electric grid is to use the existing fossil-fuel plants to fill in the gaps when renewable resources aren't available. This is almost certainly the strategy that will be employed by most utilities in the transition from fossil-fuel generation to renewables, but this strategy has significant flaws.

The most significant drawback of burning fossil fuels to provide power when wind and solar power generators are unavailable, is obvious: it requires the burning of fossil fuels. Having these older fossil-fuel generating units available will be both a blessing and a curse. The fossil-fuel generating capacity will be essential to provide energy when the renewables are unavailable but having fossil-fuel power plants available will make it too easy for system operators to call upon them whenever more power is demanded. These plants would, presumably, be operated intermittently, mostly in non-daylight hours, so at least the quantity of CO_2 released by a combination of carbon-free renewables and these fossil-fuel plants would be lower than it would have been without the renewables, but this is not a strategy to eliminate CO_2 emissions.

Solar energy production, even in the sunniest of locations, will be effectively unavailable (on average) fourteen hours a day (i.e., twelve hours at night plus one hour before sundown and another hour after sunrise when the sun's rays are impinging on the solar panels at such a low angle that power production is minimal). In the late afternoon, fossil-fuel-burning power plants will have to ramp up to meet the demand for power as solar power drops off for the day. However, because these thermal power plants are large, complicated machines, it will have been necessary to keep those plants in "hot standby" or operating at a low power level all day in order for them to be available to ramp up at sundown or whenever a cloud bank passes by. The problem is, both load-following and especially base-loaded fossil-fuel-powered units operate most efficiently at full power. Operating them in hot standby or at low power is inefficient. For the utility, operating a large power plant in hot standby would be like you sitting in a race car at a stop light for 12 hours with your engine idling so that you'll be ready to take off as soon as the light turns green. In technical terms, the "heat rate" for the plant, i.e., the total amount of fuel burned and CO_2 emitted per kWh generated, increase substantially (by anywhere from 40% to 60%) when a plant is operated at low power.

In addition, it's far better to avoid exposing large power plants (and, for that matter, most mechanical equipment) to wide temperature swings, the kind of temperature swings that occur when power plants undergo rapid changes in power output. This thermal cycling puts added stress on metal components and increases plant maintenance costs.

Option #3 - Bulk Electric Energy Storage

Option #3 in the list of how to free ourselves from fossil fuels using only renewable resources while at the same time maintaining a high level of electric system reliability is to deploy huge quantities of energy storage capacity.

To avoid having to burn fossil fuels, the idea is simple: generate excess energy from carbon-free power sources when it's available, feed that energy into a storage medium and draw it out later as needed to serve customers when the primary carbon-free generating resources are unavailable. In the case of solar energy, this means installing a sufficient number of solar panels to serve daytime energy needs, **plus** sufficient capacity in energy storage facilities to serve at least several days' worth of nighttime energy demand, **plus** enough **additional** solar capacity to generate the energy to flow into the storage systems. Most of this storage infrastructure will be owned and operated by utilities or energy supply companies, but individual homeowners or businesses, for example, could make similar investments in storage systems and the excess solar capacity to charge them.

Bulk Energy Storage Technology

There are lots of ways to store energy that have been researched and refined over the years. These include, but are certainly not limited to: flywheels, compressed air, pumped hydro, electrochemical systems (i.e., batteries), ice, hot rocks, hydrogen gas, ammonia, magnesium metal and, I suspect, dozens of others.

[*Input] Anyone who would like to offer a discussion of additional energy storage technologies beyond those discussed below for inclusion in the second edition is welcome to do so. The discussion should focus on the development status of the technology, life cycle costs compared to lithium-ion batteries, the technology's ability to be sited and any impediments to widespread commercialization such as material limitations or safety issues. [*Input –End]

Green Hydrogen

A number of researchers envision a futuristic "hydrogen economy." If you have excess non-fossil-fuel electrical energy, that energy could be diverted to an electrolysis plant that breaks the chemical bonds in water to create hydrogen gas and oxygen gas. This is effectively a storage technology because both gases can be stored and used later to produce electric energy by recombining them in a fuel cell or converting them to other energy-rich chemical compounds.

The direct use of hydrogen in natural gas distribution systems is also being tested. In one case, gas consisting of 30% hydrogen and 70% natural gas is being burned in the combustor of a gas turbine generator. In another, gas consisting of 20% hydrogen and 80% natural gas is being delivered directly to residential natural gas customers. These are not storage technologies, but they could be a way of using green hydrogen to reduce CO_2 emissions. In each application, the hydrogen displaces a portion of the natural gas so that burning the mixed gas produces less CO_2 per unit of energy obtained, if, and this is a big "if," if the hydrogen was produced without the release of CO_2.

"Green hydrogen," i.e., hydrogen produced from electrolysis powered by renewable resources, depends on two factors: 1) sufficient excess electric energy production from carbon-free sources to justify the investment in a hydrogen production and storage plant, and 2) an efficient electrolysis process. Currently neither exists.

Only about 2% (or less) of the hydrogen produced in the world today comes from electrolysis. A few percent come from coal, but roughly 95% is produced from natural gas in a process known as Steam Methane Reforming (SMR), which, unfortunately, releases CO_2 to the atmosphere. To make this process less polluting, the CO_2 released in the process could be captured and sold commercially. Such a plant exists in La Porte, Texas, where hydrogen is produced from natural gas and the CO_2 created in the process is captured, liquefied, and piped to oil fields where it's used in enhanced oil recovery (EOR) operations – not, unfortunately, a use consistent with ending our dependence on fossil fuels.

Most of the hydrogen being produced today is used to manufacture ammonia for fertilizer. Beyond fertilizer, however, there's undoubtedly a role for hydrogen in a post-fossil-fuels world, but a viable hydrogen economy remains an elusive and distant dream.

Besides developing an efficient electrolysis technology, evolving to a hydrogen economy would require the construction of dedicated facilities to produce the hydrogen. Those plants would have to be powered by carbon-free generating resources. The economics of those facilities, however, would be heavily dependent upon the capacity factors of their power sources. In other words, for a fixed amount of desired hydrogen production, it may be necessary to build a hydrogen plant that's two or three times larger than it would have to be if it were powered by a non-intermittent energy source. In addition to these production issues, widespread use of hydrogen as a primary fuel would also require an investment in an entirely new distribution network. It bears repeating that a viable hydrogen-powered economy may be an admirable goal, but it's a distant dream -- far beyond the time frame within which dramatic global warming abatement measures need to be in place. And in any case, the first step toward a "green hydrogen" future is to generate enough green hydrogen to displace the existing hydrogen production from natural gas.

Another energy storage option involves hydrogen and ammonia. Hydrogen and nitrogen can be converted to ammonia in the Haber-Bosch process: $3H_2 + N_2 \rightarrow 2NH_3$. Once produced, ammonia is more easily liquefied, stored, and transported than hydrogen and even has a higher energy density. Ammonia therefore has potential as a fuel source for combustion in specially designed gas turbines to generate electricity or as a mobile transportation fuel to replace diesel and bunker fuels in shipping applications.

However, the emergence of ammonia as a viable contributor to displacing petroleum products depends on the availability of inexpensive, green hydrogen, and as previously discussed, we're not there yet.

Low Tech Storage

There are other less exotic energy storage concepts, one of which is pumped hydro. This technology is likely to receive renewed interest in coming decades. In the age of renewables, the idea is to use excess electrical energy generated during the day to pump water from a low reservoir to a high reservoir. Then, when energy is needed to meet demand, e.g., at night, the water is allowed to flow from the upper reservoir to the lower reservoir through a turbine and in the process, convert the gravitational potential energy of the water to kinetic energy in the spinning turbine-generator. The big difference between this type of hydroelectric power generation and a conventional dam is that the working fluid, the water, is recycled between the upper and lower reservoirs over and over again, i.e., a "closed" system, as compared to a conventional hydroelectric generating station on a river where the water collects naturally behind a dam and flows through the generating equipment only once, i.e., an "open" system, as it continues to flow downstream.

As with all generation technologies, pumped hydroelectric storage has both benefits and drawbacks. [NOTE: I have to pause here. The phrase at the beginning of the previous sentence *("As with all generation technologies, . . .")* is **NOT** a casual connector. It's purposeful and it's an important concept. **Every generating technology has BOTH benefits and drawbacks.** In fact, every technology, not just electric-generating technologies, has both benefits and drawbacks. If you think you've stumbled across a technology (or a life decision) that has only "benefits," you haven't. You just haven't thought deeply enough about the technology (or the decision). One of my favorite expressions is: **"If you can't identify good reasons for NOT doing something (i.e., its "downsides"), THAT alone is reason enough to NOT do it."** Returning now to the point I was making.]

As with all generation technologies, pumped hydroelectric storage has both benefits and drawbacks. On the upside: it's simple and well developed. On the downside, siting a pumped storage plant may be problematic: there may be land and water use issues, dam safety issues and, if the water is drawn from the ocean (i.e., salt water), potential environmental concerns in and around the upper reservoir. Nevertheless, as the need for large scale energy storage capacity increases, more pumped hydroelectric projects will be proposed and many, I suspect, will be built.

Batteries Take Center Stage

In the electrochemical category, untold numbers of battery concepts have been researched for decades. These different battery types are based on different chemical reactions. One example is the lead-acid battery that's used to start the engines in our internal-combustion-engine-powered cars. Another is the lithium-ion battery that powers most modern EVs.

In recent years, a lot more funding and interest have been directed at new battery concepts. But the long history of battery research suggests that it's highly unlikely that this new research will reduce the cost of bulk energy storage by a factor of ten or even a factor of two. The biggest cost reductions are likely to be achieved through mass production. But the flip side of that coin is that mass production will drive up the cost of the raw materials that go into the more exotic modern batteries.

All battery types have a number of common issues: capital cost, maintenance cost, heat management, efficiency (i.e., what fraction of the energy charged into the battery can later be extracted to perform useful work), longevity (how many charging/discharging cycles the battery can undergo before it needs to be replaced), safety, dependence on rare or exotic materials, weight and others. Because of safety concerns, initial investment expenses and space considerations, most bulk battery storage capacity and conversion equipment will be owned by electric utility or battery storage companies.

Electrochemical bulk energy storage facilities will allow us to unlock the full decarbonization potential of solar and wind power. These facilities will look like large warehouses filled with banks of batteries, interconnected by thick cables and AC/DC conversion equipment. These facilities and the solar panels or wind turbines that will feed energy into them are the kinds of investments that weren't required when we were using storable fossil fuels. We haven't had to make these kinds of investments previously and these costs will have to be factored into future electric energy rates.

Electrochemical energy storage facilities, however, may not be as benign as large warehouses. Large battery systems can be dangerously unforgiving. They are, after all, reservoirs of large amounts of energy being prevented from escaping. (Anyone who has ever inadvertently dropped a metal wrench across the terminals of a car battery knows exactly what I'm talking about). Look no further than the industry experience with "proven" battery technologies resulting in fires and/or chemical explosions in hoverboards, laptops, vape pens, cell phones and EVs. All energy storage systems, especially massive, warehouse-sized installations, will need to be handled with care.

How Much Storage Is Enough?

If our goal is to completely disconnect from fossil-fuel-based electric generation, the required quantity of storage capacity will have to be adequate to supply the total energy needs of a utility's customers for extended periods of time – days to be sure, and perhaps weeks. In addition, the utility will have to build sufficient extra carbon-free generating capacity to be able to charge that storage capacity in a short period of time. Someone will have to pay for the construction and operation of these storage facilities and the associated renewable generation plants . . . and those investors will expect a return on their investment.

Over the next decade or two, there will be a lot of headlines about electrochemical energy storage. Many large storage projects are currently being developed and built, but **most of this energy storage capacity is not what I am referring to as bulk energy storage.** PG&E, for example, recently announced plans for multiple projects totaling nearly 720 MWe of BESS (battery energy storage system) capacity using lithium-ion batteries.

These systems typically store enough energy to deliver the rated capacity of the unit for a maximum of 4 hours. For example, a 100 MWe BESS, fully charged, would be able to deliver 100 MWe of power back to the grid for 4 hours -- 400 MWh. Alternatively, such a unit could deliver 50 MWe for 8 hours, or 25 MWe for 16 hours - in other words, diminishing increments of power for increasing lengths of time. This kind of storage **will** help the utility cope with the intermittency of solar and wind resources (by contributing to system stability and reliability) but will **not** allow fossil-fuel generating units to be shut down permanently which is the goal of decarbonization.

Stored Energy Isn't Always Green

I've identified the "Achilles heel" of solar power as its intermittency. Most authors (including myself) have identified the solution to this intermittency problem as energy storage. The idea is that energy generated and stored during the day would be fed back into the grid at night to displace the need for energy from fossil-fuel-burning power plants. That's generally the plan but adding storage capacity to the utility grid doesn't mean CO_2 emission will go down.

If any fossil-fuel energy sources are supplying energy to the grid at the time the storage medium is being charged, the energy later withdrawn from storage will be responsible for MORE CO_2 emissions than if the storage facility didn't exist. For most of the period during the transition to a decarbonized grid, storing energy generated during the day for use at night will actually **increase** GHG emissions compared to bypassing the storage medium.

This may seem counter-intuitive, but the key to understanding this conundrum is knowing that charging an energy storage medium (putting energy in) and then discharging the storage medium (taking energy out) is **not** 100% efficient. The energy efficiency of a typical pumped hydroelectric energy storage system is between 75% and 85%. The charge/discharge efficiency of a lithium-ion battery is between 90% and 95%. Because the energy storage cycle isn't 100% efficient, every unit of energy withdrawn from storage requires the production of 1.05 to 1.25 units of energy to be produced to charge the storage. Therefore, each unit of energy pulled from storage required the production of 1.05 to 1.25 times as many emissions as would have been released if the energy going into storage had instead gone directly to serve the customers' immediate energy needs or demand on the grid.

Before the era of renewables, the primary use of bulk energy storage was to reduce energy costs by using generating capacity with a low marginal operating cost or by eliminating the need to build additional generating capacity. For example, a generating resource with a low marginal operating cost (e.g., a nuclear power plant) would be used to charge a storage facility (e.g., a two-reservoir pumped storage hydroelectric facility) at night when grid demand was low. Energy would be withdrawn from the storage to serve peak demand during the day with energy that had a lower net cost of production than alternatives, even accounting for the efficiency loss associated with the storage process. Alternatively, if fossil fuels were burned to generate the energy to charge the storage facility, net energy costs would still be lower because the cost per kW to build the storage facility would be lower than building an additional fossil-fuel power plant.

In the era of renewables with the objective of reducing CO_2 emissions, the calculation is different. **Every kWh of carbon-free energy (solar, wind, nuclear, geothermal, etc.) that displaces the need to burn fossil fuels reduces total CO_2 emissions.** However, if fossil fuels are being burned to supply the grid when the storage facility is being charged, the CO_2 reduction benefit of the carbon-free resource is reduced by the inefficiency of the charging/discharging cycle.

The top two curves in Exhibit 9.1 reproduce the grid power demand on a system with and without a substantial contribution of solar energy (i.e., Exhibit 8.4). The additional curves in this exhibit show the demand on the grid as progressively more solar capacity is added. The bottom curve represents the case where so much solar capacity has been added to the grid that all midday fossil fuel production has been displaced by solar energy and excess energy is available to be diverted to storage with the maximum CO_2 reduction benefit. (Note: this discussion ignores the presence of other carbon-

Exhibit 9.1
Grid Demand As Progressively
More Solar Capacity Is Added

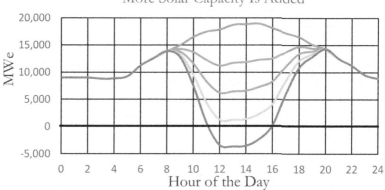

free resources on the grid such as wind, nuclear and geothermal power which would cause the "excess solar capacity condition" to be achieved with the addition of fewer solar resources. This example also ignores the operational complications and CO_2 emissions implications of cycling large fossil-fuel generating stations or operating them in "hot standby" as discussed in Option #2 earlier in this chapter. In either case, the conclusions are the same). Charging storage capacity with fossil fuel generated power increases emissions compared to what they would have been without the storage. Charging storage with energy from carbon-free resources before the grid has been fully decarbonized reduces emissions, but by a lesser amount (a storage inefficiency penalty) than would have been realized if that energy were used to serve an instantaneous demand. Homeowners with solar systems seeking to maximize CO_2 emissions reductions may want to forego energy storage systems for the foreseeable future.. Exhibit 9.1 also illustrates why the optimal utilization of carbon-free energy for storage may be limited to a few hours per day between 10 am and 4 pm.

Option #4 - The Combo Compromise

A combination of all three strategies will be used to serve power demand in the era of renewables:

1. An intentional (slight) reduction in the reliability of the electric utility grid,
2. The burning of fossil fuels to supply energy when renewables cannot, and
3. A massive deployment of energy storage systems.

Even assuming that the deployment of renewable resources exceeds all expectations, we will still need redundant power generating equipment and storage capacity. Ideally, the renewables power option will provide the majority of our electric energy needs, especially during the day, and the fossil-fuel-powered option will only be used when solar and wind resources aren't available. Having redundant power generating resource, however, will be expensive.

If only there were another way . . .

An Alternative to Energy Storage

Up to this point, my discussion of energy storage has been based on our society's current pattern of energy consumption. That pattern was shown in Exhibits 6.2 and 8.3.

An alternative to storing energy produced at one time for use at a later time is to shift that energy-consuming activity to a time when energy is available, thereby reducing the need for storage. For example, rather than running a dishwasher in the evening after sundown and powering it with solar energy collected and stored in a battery during the day or operating a fossil-fuel-powered generating plant, an alternative approach would be to run the dishwasher during the day when excess energy is available from roof-mounted solar panels or from the grid. This would diminish the amount of storage capacity that would have to be deployed either by a homeowner or an electric utility. (Depending upon other daytime energy demands in the household, this approach might require a homeowner to install a greater number of solar panels).

Dishwashing is just one example of an energy-consuming activity performed typically in the evening that could be shifted to a time when energy from intermittent resources is more available. In addition to dishwashing, laundry washing and drying, some types of cooking, food refrigeration, space heating and space cooling (i.e., air conditioning) could all be "time-shifted" in whole or in part from time periods when solar energy is unavailable (after sundown) to times when solar energy is typically more abundant (after sunrise).

Another approach to time-shifting energy consuming activities is to use ice storage systems in commercial buildings. In these systems, water is frozen during periods when electric energy is available and less expensive and later thawed to provide cooling when energy is more expensive. Refrigerators with the built-in ability to store "cold" (usually internal systems to freeze water or another fluid) have been demonstrated. Refrigerators with this capability could remain adequately cool without running the refrigerator's compressor at night.

California farmers have been "time-shifting" crop harvests for a long time. In the heat of summer, some farmers send their laborers out into the fields after midnight (long after the heat of the day has subsided). They use floodlights to illuminate the work areas (which, of course, uses energy), but the workers have a much lower risk of heat exhaustion, productivity is higher, and the crops arrive at the processing plant cooler than they would have if they had been picked during the heat of the day.

Most of the time-shifting energy systems and policies in effect today were developed for reasons unrelated to CO_2 emissions. These time-shifting energy systems were developed to move energy consumption from daytime hours when energy demand was greatest to nighttime hours when total demand from the grid was much lower.

For years, our local utility, SCE, ran an advertising campaign to "Give Your Appliances the Afternoon Off." These public service spots ran prominently in the summer months when the regional air conditioning load was the greatest. Participation was voluntary. The company was attempting to reduce the peak demand on the electrical grid by appealing to customers to defer discretionary energy use to a period in the day, either in the morning or evening, when demand was lower. The motivation for the company and its customers was that the company wouldn't have to purchase as much higher priced energy from neighboring utilities, build more power plants or operate its own more expensive peaking units. This helped keep utility operating costs down and electric bills from rising as fast.

As more solar power is added to the grid in Southern California and elsewhere, this voluntary shifting of discretionary energy use is being turned on its head. With excess solar generation available during daylight hours, the utility is appealing to its customers to "Give Your Appliances the Evening Off."

SCE is currently running public service announcements discouraging energy use from 4:00 pm to 9:00 pm (Chapter 10). This effort to get customers to voluntarily curtail energy use in the late afternoon and early evening is a direct result of trying to accommodate large quantities of solar power generation on its grid. Solar generation drops rapidly in the late afternoon, but power demand persists into the early evening. By asking its customers to curtail energy use in this period the utility is trying to avoid having to purchase large blocks of more expensive fossil-fuel-generated energy.

This is a logical and appropriate response by the utility, but they're missing an opportunity to help their customers transition more fully into the era of renewables. In addition to urging customers to reduce their late afternoon energy consumption, they should also be encouraging customers to perform energy-intensive activities in mid-day on sunny days when solar energy is most plentiful.

In addition to public service announcements, another tool available to utilities to tailor energy consumption patterns is pricing, specifically: Time-of-Use (TOU) rates (cost per kWh), interruptible rates and other load-shaping incentives. TOU rates are lower during periods when energy is readily available and the cost of production is low. Conversely, TOU rates are higher when energy is less available and more costly to produce. Interruptible rates provide customers with a rebate if they agree to allow the utility to briefly turn off a specific appliance (typically a central air conditioning system) when power demand is high or energy is expensive.

For example, I've signed up for a program that allows the utility to turn off my central air conditioning (i.e., their choice, not mine) between noon and 6 pm during the hottest months of the year on days when power supplies are tight and/or energy is expensive. In return the utility gives me a rebate for the months of the year when the program is in effect. In my case, however, allowing the utility to cut the power to my central air conditioning system hasn't resulted in any loss of comfort. I've insulated my home to the max and on days when the temperature is predicted to be uncomfortably high, I run my air conditioning system for an hour or two in the early morning to pre-chill and dehumidify my home. I manually turn off the air conditioning system long before the time when I've given the utility the authority to cut the power to my system automatically. The program ensures that my air conditioning load will never add to the peak demand or higher energy acquisition costs --and in return, I get a welcome rebate.

The same kind of approach can be used in cold climates where house designs could incorporate a thermal mass (e.g., tanks of water, extra masonry, volumes of rock, volumes of sand, etc.) which is heated at one time of the day when energy prices are low and from which heat (energy) is extracted at another time of the day when energy supplies are not available or are expensive.

Many modern appliances now have the capability to program the start-time of the appliance. "Programmable thermostats" have been doing this for years for central heating and air conditioning systems. In addition, homes are increasingly being equipped with control systems to turn on and turn off any number of appliances automatically. These systems should provide owners with the capability to tailor their energy consumption pattern to take advantage of lower electricity rates. "Smart" appliances could even interrogate the utility grid to determine the current energy price and turn themselves on or off based on a cost algorithm specified by the customer.

The goal of the earlier energy consumption time-shifting strategies and incentives (public service announcements, specialized rates, rebates on new equipment, etc.) was exactly the opposite of what we will need to do in the era of renewables. If the goal today is to minimize the amount of energy storage capacity we have to build and operate, then these energy-consumption time-shifting programs will seek to shift energy consumption from nighttime hours to daytime.

Ultimately utilities will almost certainly transition to dynamic energy pricing where the price of energy will increase or decrease on a moment-by-moment basis throughout the day based on the cost of production and availability of energy. Dynamic energy pricing will be the most equitable for all consumers and provide an incentive for energy-conscious individuals to actively manage their energy usage and electric bills.

It's important to note that changing the time of the day or the day of the week when an energy-consuming activity is performed **doesn't reduce** the amount of energy required to perform the task. It has the potential to decrease only the amount of energy storage capacity that will have to be deployed. Nevertheless, avoiding the requirement to deploy additional energy storage may be of a significant benefit to customers of the future electric grid.

Winding the Spring Tighter and Tighter

Time-shifting to avoid building even more energy storage capacity and using an energy storage strategy to decrease the use of fossil fuels introduces several operational complications. If we succeed in moving significant energy consumption from nighttime to daytime, we may suffer even more severe generation constraints at those times when solar power is unavailable. During prolonged cloudy periods in regions heavily dependent on solar resources, we'll face higher energy prices, power shortages, and extended blackouts until the skies clear, especially if we consumers have been conditioned to use energy during the day.

In addition to time-of-the-day energy shifting, strategies may also develop to shift energy usage from certain days when energy consumption is greater to days when demand is lower. Energy consuming activities may be ranked on the basis of their importance. In the event of a multi-day period when solar power is unavailable, utilities and consumers may choose to delay some less critical, energy consuming activities for several days (rather than just several hours) until a time when the renewable resources are available again. For example, a consumer could delay doing the laundry for a day or two during a cloudy period.

Time-shifting energy consumption patterns will not dramatically alter the dangerous trajectory we're on in regard to global warming. I've discussed them here as examples of how our lives will change in the era of carbon-free generating resources and that's really the underlying message of this book: to decarbonize our energy economy we'll have to rethink everything we do and how we do it and even as we do, the sad truth is that sometimes we'll be unable to meet demand without burning fossil fuels.

Even Better

Another way of reducing the amount of energy that may need to be stored is to move certain energy-consuming activities off the grid entirely. Clothes drying is an example. My mom, a farm girl in her youth, dried clothes on a clothesline long after she left the farm and even after we acquired a clothes dryer. (For good or bad, habits and values acquired early in life often have a life of their own). The convenience of modern, clothes dryers powered by fossil fuels has pretty much forced the clothesline to go the way of the buggy whip, at least in most of the U.S. Around the world, however, clotheslines are commonplace. As energy becomes more expensive and efforts to reduce CO_2 emissions really take hold, I predict a widespread return of the clothesline in the U.S. – perhaps not in my lifetime, but they will, like nuclear power plants, make a comeback.

Looking Ahead

Even though the "fuel" for solar panels and wind turbines is free, the cost of energy from a grid heavily dependent upon intermittent, renewable resources will be more expensive than its fossil-fuel-burning predecessor. This is because of the need for bulk energy storage capacity, additional resources to recharge that storage capacity, additional transmission lines and the residual expenses for underutilized fossil-fuel generating capacity.

Please note: I'm NOT arguing we shouldn't make this transition to carbon-free resources because of their higher cost. We absolutely should. I'm simply explaining some of the reasons why the energy from the carbon-free grid of the future will be (much) more expensive. It will be cleaner, yes, but it will be more expensive. A "carbon-free" energy future won't just happen and it won't be cheap.

> *A "carbon-free" energy future won't just happen, and it won't be cheap.*

Action Items

As relates to energy storage, here are a few of the projects and policy initiatives that could help address global-warming-related stresses.

1. Identify sites for and build large, pumped storage hydro-electric generating projects on an accelerated basis.
2. Build more low-head, flow-of-the-river power generating plants.
3. Direct utility regulators to set modestly lower reliability standards for electric systems.

10 THE FUTURE GRID

"It's tough to make predictions, especially about the future."

An old Danish proverb (Popularized by and often attributed to Yogi Berra)

Our current electrical grid is highly efficient. It predominantly uses fossil fuels that have a high energy density, are easily stored and transported, and are comparatively inexpensive. For the last 100+ years, we've been enjoying the low hanging (energy) fruit we call fossil fuels.

> *For the last 100+ years, we've been enjoying the low-hanging-fruit. Future generations will have to work a lot harder, i.e., pay a lot more, for their energy.*

What we're now talking about is transitioning to an energy grid that utilizes energy assets (the sun's rays and moving air) that may be free, but are diffuse (dilute, not concentrated), intermittent, and, in their raw form, can't be stored. Future generations will have to work a lot harder, i.e., **pay a lot more**, for their energy than we have in the era of fossil fuels. Of all the peoples who have ever lived on earth, just three or four generations, i.e., our parents, us and our kids have enjoyed the benefits of cheap energy. We've created a world energy infrastructure and a global economy that are completely defined by and fully invested in cheap fossil fuels. The CO_2-generating spigot is wide open and won't be easy to shut off. Before the words "global warming" ever passed anyone's lips, we started down the path we are on today and that path is like a superhighway with no easy off ramps.

By now I hope my vision of the future electric supply grid is coming into focus. There will be no silver bullet that's going to save us and no single formula to free ourselves from fossil fuels. A variety of technologies and policies to supply energy and minimize the consumption of energy will all be utilized to reduce

> *No silver bullet and no single formula for success will free us from fossil fuels. A variety of supply-side and demand-side technologies and policies will all be utilized.*

our long-term CO_2 emissions. As energy prices rise, a greater variety of both supply and demand side options will come into play.

Solar and wind energy generation are now, and will continue to be, the most cost-effective, highly developed renewable resources. They're likely to provide most of the new carbon-free energy in the U.S. and around the world for the next several decades. Some regions will deploy these resources at a rapid rate based on their geographic location and the economics of the resource. Nationally, however, the transition in the U.S. away from fossil fuels is likely to be slow. The most difficult part of the transition will involve the hours of the day when solar power is unavailable. Every fossil-fuel-generated kWh displaced by a carbon-free resource is a step in the right direction, but unless enormous quantities of bulk energy storage capacity are installed, fossil-fuel-powered electric generation will continue to be required in the evening and nighttime hours or at other times when solar output is unavailable. The good news is that these fossil-fuel-powered generating resources are already in place. The bad news is that operating them in a demand-following mode will be inefficient, hard on the equipment, very expensive and of course . . . will continue the emission of CO_2.

The fact that these coal and natural gas generating stations are already in place has another downside. The mere fact that they're available will make it too easy to use them. If the installation of renewable resources and storage capacity falls short of what's needed, the owners of the fossil-fuel-generating stations will be more than willing to step in to save the day. The additional threat is that when people become fed up with the changes demanded of them to meet certain "green" energy targets (and they will at times get fed up), they may insist that the fossil-fuel plants be fired up.

The future grid will have an enormous quantity of energy storage capacity. The investment to fulfill that requirement to make the grid green will also be enormous. Sufficient storage will have to be installed to provide energy when solar power is unavailable or diminished due to seasonal variations. To fully disconnect from fossil-fuel generation **and** maintain the same level of system reliability (or even

> *The key to a renewable energy future is not only renewable energy generation capacity, it's also energy storage capacity and the renewables to charge that storage.*

close to it) the local utility might have to build carbon-free generating resources equal to several times the total amount of its fossil-fuel generating capacity and enough energy storage to store as much energy as its fossil-fuel power plants supply to the grid over a multi-day period. The number of days this storage capacity might be called upon to power the grid without a fossil fuel backup will determine the sizing of this energy storage capacity. The investments in this storage capacity will be enormous and virtually none of it currently exists.

Prior to the introduction of significant quantities of solar and wind generating resources, the utilities in various parts of the country looked fairly similar in terms of their electric generating fuel mix. Some areas had more hydroelectric or nuclear capacity than others, but most utilities relied heavily on natural gas and coal. The split between coal and natural gas was largely determined by the cost and proximity of the fuel, air quality requirements and gas pipeline capacity.

Looking forward, utilities around the country are likely to look less similar. Utilities will utilize solar power most heavily in the Southwest as suggested by Exhibit 8.1. In solar-rich areas, utilities could have more solar generation than they can use in the milder months (typically the late fall, late winter, and spring). When these conditions develop, large quantities of energy will be transported over the transmission system to areas of need. At these times, utilities will also pay customers with solar panels less for the energy they're supplying to the grid because the supply will be comparatively abundant, and demand will be comparatively low. This excess solar generation during the day in these shoulder months will also drag down the price paid to wind power producers. Regions other than the Southwest may rely more heavily on wind power which may be easier for utilities to manage because of its higher capacity factor and potential availability at night.

Energy pricing policies will be used to send ever stronger behavioral modification signals in the future. Customers (residential, commercial, and industrial) will respond with a wide variety of new equipment and appliances, plant improvements, process changes and lifestyle choices. Some of the need for storage capacity will be offset by time-shifting a range of demand activities and through conservation. This will be a slow, expensive transition. Because it will be slow, our planet's CO_2 burden will continue to grow. Because it will be expensive, many forces will resist and try to delay the changes.

And Now the Numbers

I've described the challenges associated with disconnecting from our current fossil-fuel electric generating grid and switching to renewables. I've characterized the two most prominent renewables, wind and solar, as intermittent in nature, requiring the installation of massive quantities of energy storage (after the grid has been decarbonized) to ride through periods when those intermittent primary sources are unavailable. The investment in carbon-free resources to supply energy during the day and the storage capacity to supply energy when the intermittent resources are unavailable will only be operationally beneficial if massive quantities of wind and solar resources to recharge that storage capacity are also installed and substantial investments are made in our high voltage transmission system.

Exhibit 4.1 showed that the U.S. was estimated to have consumed just over 100 Quads of energy in 2019. Of that total, the useful electric generation portion was equivalent to approximately 12.7 Quads. Of that, coal, natural gas and a minuscule amount of petroleum were used to produce approximately 7.4 Quads of useful electrical energy. If you wanted to repower all 7.4 Quads of fossil-fuel-powered useful electric energy with rooftop solar panels mounted on people's rooftops (**you would never do that**, but to just get a feeling for the magnitude of the undertaking) you'd have to install **4.7 billion solar panels**. (1 Quad $=1 \times 10^{15}$ BTUs, 1 kWh $=3,412$ BTU, a typical solar panel $= 0.33$ kW, typical residential solar panel capacity factor $= 16\%$, 1 year $=8760$ hours. (A typical household solar installation consists of between 20 and 30 panels).

This calculation assumes residential rooftop solar installations have a capacity factor of around 16%. As previously discussed, utility owned and operated solar installations have a higher capacity factor, roughly 25%. If future solar installations were a 50/50 mix of residential and utility projects, the above numbers could drop by roughly 20%.

Some researchers have raised the alarm that manufacturing massive quantities of solar panels (and batteries) would put a premium on some exotic minerals used in their production. Aside from the special minerals involved, it seems reasonable to assume that if you produce billions of anything, you're likely to run up against material, labor, manufacturing, or capital constraints of some kind.

Similarly, if you wanted to repower the 7.4 Quads of fossil-fuel-generated energy with wind turbines alone **(again, you would never do that,** but to just get a feeling for the magnitude of the undertaking), you'd have to deploy **287,000 medium-size modern wind turbines**. (2.5 MWe per wind turbine, capacity factor 34%)

It should be noted that the above calculations assume that all the currently operating, carbon-free resources continue in operation i.e., adequate water flows on dammed rivers to maintain current levels of hydroelectric power, continued operation of all current nuclear, solar and wind resources. These calculations also assume that future solar and wind installations are as productive as current units, but future deployments of solar panels will increasingly be made in less-favorable, more-northerly, and less sunny locations. Similarly, future wind turbine installations will certainly seek to exploit slightly less favorable wind resources (because the best ones would have already been developed).

These calculations of the number of carbon-free generating resources necessary to displace fossil-fuel consumption on the national electrical grid do not account for population growth. In 2019, the U.S. population increased by approximately 0.6%. (That growth rate is expected to trend down to around 0.4% by 2050). If the per capita electric energy consumption remains the same as it is today, population growth alone will require an additional 1.8 Quads of carbon-free electric energy per year by 2050. In other words, population growth could increase the above numbers of new solar panels or wind turbines needed to achieve a carbon-free electric grid in 2050 by 20%.

These numbers do include the renewable resources to produce the energy that's diverted to storage during the day for use at night daily. They do not, however, include the additional renewable resources that would have to be deployed to charge the storage capacity that would allow a region to continue to be served with stored electrical energy during extended periods when power production from renewables was not available.

These numbers also don't include the numbers of new solar panels or wind turbines needed to decarbonize the 25.8 Quads of petroleum energy consumed in the transportation sector (to be discussed in Chapter 11) or the 20.4 Quads of natural gas energy delivered directly to homes, businesses, and industry (Chapter 12).

Our electric energy grid of the future will have enormous quantities of solar and wind generation, but we will not even come close to deploying the billions of solar panels or hundreds of thousands of wind turbines or the quantity of energy storage capacity and new transmission lines necessary to **disconnect completely from fossil fuels**. I can readily envision a day when a combination of wind and solar power supply 50% or more of the country's (and the world's) daytime electric energy demand and storage and wind power contribute 30% or 40% of the nation's nighttime energy needs. If we achieve this level of dependence on renewables and storage, we will still not be able to turn off the fossil-fuel spigot feeding into our power plants. Without a carbon-free, dispatchable (i.e., non-intermittent, utility-controlled) power source such as a nuclear plant, and lacking sufficient energy storage capacity, we'll have to continue operating a significant fraction of the fossil-fuel-based power grid to maintain system reliability for the foreseeable future. These power plants will, of course, continue emitting CO_2.

Giving Nuclear Power a Second Look

I've already made the point that public opposition to nuclear power, or at least a lack of widespread enthusiastic support, will preclude this power generating option from contributing to combating global warming in any meaningful way until well past 2050. If, on the other hand, nuclear power was utilized, the entire fossil-fuel component of the existing retail U.S. electrical grid could be displaced by fewer than **263 large nuclear power plants.** We currently have around 93 nuclear reactors operating on 55 sites around the country. With two reactors per site, nearly the entire fossil-fuel electric generating capacity in the U.S. could be replaced by carbon-free nuclear reactors operating on fewer than 130 sites around the country

But that's not going to happen. Nuclear power could help mitigate the global warming imbalance, but too many people have a strongly negative perception about a technology they know very little about. They may think it's unsafe or too expensive but recall that at one time nuclear power was safely producing over **three-quarters** of the electrical energy consumed in an entire country (i.e., France) – all carbon-free. But somehow the deployment of nuclear power plants is scarier to a significant fraction of the population than a vision of the **earth turning slowly on its axis being roasted like a rotisserie chicken**.

There is one other characteristic of nuclear power plants that strongly suggests that nuclear power plants could play a prominent role in displacing CO_2 emissions in the latter half of this century. Nuclear plants are most efficient and best suited to operation as base loaded generation capacity. These are the plants that operate 24/7, 365 days a year. It turns out that the combined power generating profiles of two technologies, solar and nuclear, are actually a pretty close approximation of the resources necessary to satisfy a typical electric power demand.

The nuclear plants would operate day and night, serving the base load during both the day and night while the solar contribution to the grid ramps up along with customer demand as the sun rises, peaks out in mid-day when customer consumption is highest and tapers off along with demand into the evening.

Exhibit 10.1 shows the daily power production profile of a 1000 MWe operating nuclear plant, a pleasantly boring and constant 1000 MWe all day long.

Exhibit 8.2 showed the daily power production profile for 1000 MWe of solar capacity. Combining the power production profiles of solar and nuclear (Exhibits 8.2 and 10.1) and scaling up to match simulated power demands in Southern California) produces a combined power supply profile like that shown in Exhibit 10.2. Also shown in Exhibit 10.2 is a typical summer power demand profile. Exhibit 10.3 shows the comparable profiles for a typical winter day.

Exhibit 10.1
1,000 MWe Nuclear Output by Hour of the Day

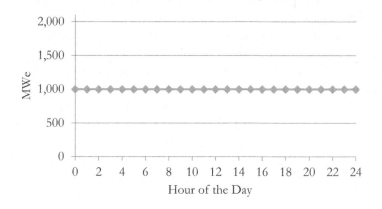

Exhibit 10.2
A "Marriage" of Solar and Nuclear
to Serve the CA Summer Demand

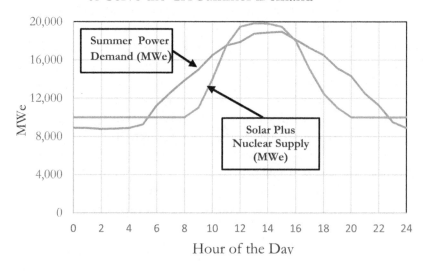

Exhibit 10.3
A "Marriage" of Solar and Nuclear Power
to Serve the CA Winter Demand

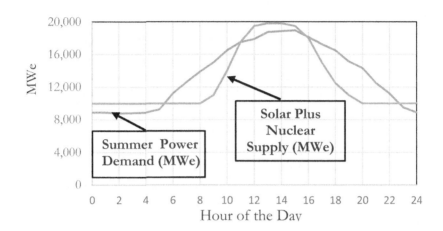

You don't have to squint too hard to see a close match in both Exhibits 10.2 and 10.3 between the shape of the total daily power usage profiles and a power supply profile consisting of a mixture of solar and nuclear generating resources. The other benefit of this energy technology "marriage" that doesn't immediately jump off the page is that it dramatically reduces the required amount of energy storage capacity and renewable resources to recharge that storage.

A full-scale deployment of carbon-free solar, wind and nuclear generating capacity wouldn't completely eliminate the need for energy storage and additional generating capacity to charge that storage, but the quantity of storage and the cost of that investment will be a tiny fraction of what it would have been without the nuclear component.

When the logistical and financial challenge of deploying ONLY solar, wind and energy storage capacity resources to power a reliable power grid finally become apparent, public opposition to nuclear power will shift. This WILL happen, but long after GHG concentrations in the atmosphere drive many unpleasant outcomes.

11 TRANSPORTATION

"You can't get there from here."

Expression attributed to residents of Maine.

The photos below are four images taken from a freeway overpass near my home (at the height of the pandemic 2020, i.e., exceedingly "light" traffic!): one each at 7:00 am, noon, 6:00 pm and 11:00 pm on the same day.

Exhibit 11.1 Freeway 7:00 am   Exhibit 11.2 Freeway Noon

Exhibit 11.3 Freeway 6:00 pm   Exhibit 11.4 Freeway 11:00 pm

There's nothing particularly remarkable about these photos. In fact, the amazing thing about these images is that they're so unspectacular, so familiar. Admittedly, you won't see images like this in rural Kansas, but these images reflect the daily lives of so many people around the world.

The Real Cost of Going from Point A to Point B

Excluding aviation fuel, in 2018 Americans consumed almost **400 million gallons of gasoline and diesel fuel PER DAY**. That's an average of more than a gallon per day for every man, woman and child in the country. Worldwide oil production is somewhere between **3.5 and 4.2 billion gallons per day**. If you think about it, this is a really astounding number. Consider just the logistics of extracting that quantity of product from deep in the earth, transporting it to refineries, processing the raw petroleum into a range of products, transporting those products to points-of-sale, and finally distributing them to customers every single day, 365 days a year. It's unfathomable. The logistical accomplishment of distributing four billion gallons of refined petroleum products every day would be something to celebrate in amazement if it weren't killing the planet.

The logistical accomplishment of processing four billion gallons of refined petroleum products per day would be something to celebrate in amazement . . . if it weren't
KILLING THE PLANET.

Exhibit 4.1 showed the estimated sources and uses of energy consumption in the U.S. (2019). In that figure, the massive black bar on the bottom of the graphic represents the energy consumption from petroleum. Excluding natural gas, the energy content of the petroleum consumed by the U.S. is greater than all the other sources of energy combined: greater than all the (retail) solar (electric), nuclear, hydro, wind, geothermal, coal and biomass, **COMBINED**. Some petroleum is used in industry, but transportation accounts for the lion's share.

CO_2 emissions attributable to the transportation sector in the U.S. are roughly the same as the emissions from the generation of electricity. If we have any hope of "beating" global warming, we can't just deploy wind turbines, solar panels, and energy storage facilities to "green" the electrical grid. We will also have to focus on cars, trucks, trains, planes, ships, and everything else that makes up the transportation sector.

More than 80% of worldwide petroleum production is refined into transportation fuels: gasoline, diesel, jet fuel and heavy bunker fuels (most commonly burned on ships). Comparatively small quantities of natural gas are also compressed

> **The EPA estimates that 29% of CO_2 emissions in the U.S. are attributable to the transportation sector, comparable to the emissions from the generation of electricity.**

and used as a transportation fuel. Of these fuels, the vast majority of liquid petroleum products are used in personal automobiles and light trucks.

Some transportation applications such as mass transit systems (subways, light rail, etc.) and electrified heavy rail systems use electric energy straight off the grid, but the majority of transportation needs are powered by petroleum products.

Like the combustion of all fossil fuels in electric power plants, one of the by-products of the combustion of gasoline and natural gas in an internal combustion engine is CO_2. If "beating" global warming requires that we stop emitting CO_2 into the atmosphere, we'll have to replace the petroleum-based transportation fuels with a practical, reliable alternative form of mobile, stored energy.

Alternative Transportation Fuels

One possibility would be to switch to agriculturally derived fuels like ethanol. The one favorable dimension of using agricultural output to produce a liquid fuel is that its combustion wouldn't add "new" CO_2 to the atmosphere. Plants extract CO_2 from the atmosphere and then the ethanol made for those plants would be returned to the atmosphere when the liquid fuel is burned.

Ethanol, however, is not "the" solution that will yield a carbon-free transportation sector. The diversion of agricultural output from the production of food to the production of transportation-ethanol is already controversial and will become increasingly so in the future as worldwide food demand increases. However, beyond that, even if ALL of the corn grown in the U.S. were diverted to the production of ethanol, it would displace just around 12% of all gasoline consumed in the country. In fact, it would take almost twice as much land as is currently used to grow ALL crops in the U.S. to produce enough corn for processing into ethanol to match our current gasoline consumption. Nobody is talking about doing that of course, but it does give you an idea of the enormity of the challenge to repower the entire petroleum-based transportation sector.

As the world's population continues to increase and as historically productive acreage become less productive due to extreme heat and/or flooding, food supplies are likely to become a more urgent concern in the not-too-distant future. Some regions could become generally more productive than they are now due to a warming climate, but even those areas could be subject to crop-damaging extreme heat events. Fortunately, as other viable power sources for transportation (such as electrochemical batteries, hydrogen and synthetically manufactured liquid fuels) are developed further and become more readily available, the rationale for using agricultural products to produce liquid fuels for transportation will fade away. In the not-too-distant future, farm products will revert to their historical usage as food for hungry people.

It's possible to manufacture liquid fuels from readily available feedstocks including water, CO_2, and carbonic acid. Doing so, however, requires one additional crucial ingredient – a cheap, high temperature energy source. Chemists and chemical engineers are modern day "alchemists." They can make just about anything by rearranging chemical combinations and structures, but even they are powerless to do any good without a carbon-free, high temperature energy source to drive chemical reactions in reverse.

There is one such high temperature, carbon-free energy technology which I've already mentioned in connection with electric generation and that's nuclear power. There are some advanced nuclear reactor concepts that can drive a gas turbine and/or steam cycle turbine to produce electricity but can also deliver high temperature process heat to drive chemical processes. However, if the "conventional" nuclear option is unavailable to be used to generate electricity for the reasons I've already discussed, the advanced nuclear concepts to deliver process heat, which will require additional development, testing, infrastructure and deployment, are even that much further out of reach, at least until after 2050 or later.

A limited number of hydrogen-powered, fuel cell vehicles are on the road and are continuing to be tested. For the most part, these vehicles still rely on fossil fuels (i.e., hydrogen produced from natural gas) and therefore, contribute to CO_2 emissions.

[*Input] – Anyone with first-hand knowledge of non-battery, non-fossil-fuel-powered personal mobility options (not bicycles, thank you) is invited to submit an article for inclusion in the 2nd edition. The article should specify how the fuel source is derived, the development status of the technology, how the economics of the option compare to conventional battery powered EVs, the scalability of the technology and the infrastructure requirements for widespread deployment. [*Input - End]

EVs to the Rescue?

For now, at least, electric vehicles (EVs) are the leading contender to replace gasoline and diesel-powered cars and light trucks. In order for EVs to make a substantial contribution to reducing GHG emissions two things have to happen.

First, EVs need to become a much greater share of the vehicles on the road and that transition needs to happen fast. Currently there are more than **a billion** petroleum-powered vehicles on roads around the world. There are over 270 million registered vehicles in the U.S. alone. Each year, worldwide car and light truck sales exceed 77 million units; in 2019, U.S. car sales exceeded an astounding 17 million vehicles.

In 2019, plug-in hybrids and battery-only EVs reportedly accounted for between 2.5% and 4% of vehicle sales (the wide range in these data seems to be the result of differing definitions of what constitutes an EV). The percentage of annual EV sales in the U.S. will continue to rise, and the percentage of EVs on the road across the U.S. and around the world will also rise. EV manufacturers will be competing in an increasingly crowded market space which, until fairly recently, was dominated in the U.S. by just one manufacturer. More suppliers will be good for consumers.

With clean vehicle purchase incentives and/or increased gasoline taxes, it's possible that EV sales will increase to 25% of all new vehicle sales by 2035 or 2040, which would be a dramatic development. But still, these numbers are too small to make a significant difference in CO_2 emissions attributable to the transportation sector in the next two or three decades. This is partly due to the fact that the average life of a car on the road in the U.S. is 10 to 15 years. In other words, the vast majority of gasoline-powered vehicles rolling of the assembly line today will still be on the road in 2035. Even with a dramatic increase in the EV share of the new car market, the majority of the cars on the road in the U.S. and around the world in 2050, and the majority of vehicle-miles per year will still be powered by petroleum.

The second thing that has to happen for EVs to make a significant dent in transportation-related emissions takes us back to the electrical grid. Unfortunately, electric vehicles are probably not the environmental panacea that many people think they are. Electric vehicles are advertised as "zero-emission" vehicles, but that characterization is more than a little bit **misleading.**

> *EVs are probably **NOT** the environmental panacea that many people think they are. EVs are only as "clean" as the grid that recharges their batteries.*

(A friend convinced me to use the word "misleading" in place of my original word choice, **"dishonest."**)

Electric vehicles may, more accurately, be characterized as "zero-tail-pipe-emissions" vehicles. The energy to recharge an EV's batteries still has to come from somewhere. All of the emissions produced at the power generating facility to generate the energy to recharge an electric vehicle's batteries have to be added into the calculation.

Consider the energy consumption associated with running an EV recharged off the grid where natural gas is the marginal generating fuel. I'm using natural gas in this example because it's the cleanest of the fossil fuels. If the EV consumes about 0.25 kWh per mile driven, the battery charging efficiency is around 90%, and distribution line losses are 2%, then it takes about 0.28 kWh generated at the power plant to recharge an EV to drive a mile. (Note: these numbers are characteristic of a large EV like a Tesla. Smaller, lighter EVs may consume less energy per mile, but at least for now, Teslas are the most common EV in California). The conversion factor to express the energy content of 1 kWh in BTUs (British Thermal Units) is 3412 BTUs/kWh. Therefore, the energy content of the 0.28 kWh delivered to the EV owner's home is equal to around 955.4 BTUs of (electrical energy equivalent) at the power plant. However, a typical natural gas power plant operates with a thermal efficiency of approximately 33%. Therefore, in order to generate 955.4 BTUs of useful electrical energy for an EV to drive a mile, the power plant had to consume (i.e., burn) enough natural gas to yield 2,866 BTUs. In other words, the EV actually consumes roughly 2,866 BTUs of the natural gas source fuel to travel one mile.

This energy consumption of an EV can be compared to the energy consumption for a high mileage gasoline powered vehicle. The energy content of one gallon of gasoline is 124,000 BTUs. A **high efficiency** gasoline-powered vehicle might get in the neighborhood of 45 miles per gallon (mpg), (i.e., far higher than the average for the existing U.S. vehicle fleet which is around 22 mpg). **The high mpg gasoline-powered vehicle**, therefore, **consumes** about 2755 BTU per mile, comparable to, but actually **less than the most popular EV** (recharged off the grid).

Energy consumption is one dimension of performance, but for the purposes of this discussion of global warming, we also need to look at EVs from the perspective of CO_2 emissions. (If the source energy consumption per mile for the two types of vehicles is comparable, as described above, you can imagine that the CO_2 emissions for the two are also similar and you'd be right). Burning gasoline generates approximately 19.64 pounds of CO_2 per gallon consumed. The gasoline-powered vehicle in this example (a 45-mpg vehicle) therefore, emits around 0.436 pounds of CO_2 per mile. The source fuel for charging the EV battery in this example is natural gas which produces about 1.21 pounds of CO_2 per kWh. The EV in this example is therefore responsible for roughly 0.34 pounds of CO_2 emissions per mile, only about 22% better (lower) than today's high mpg gasoline-powered vehicles. In other words, the typical EV is roughly equivalent to a high mpg gasoline-powered vehicle in terms of energy consumption and only marginally better in terms of CO_2 emissions. In any case, **an EV charged off the grid is most definitely not a "zero" CO_2 emissions vehicle.**

The above analysis assumes that natural gas is the fossil fuel utilities are burning to produce the energy

A large EV produces only marginally less CO_2 than an efficient (high MPG) gasoline-powered car.

that flows into the EV batteries. A similar analysis can be performed using coal as the fossil-fuel source for generating that energy. Natural gas produces roughly 1.21 pounds of CO_2 per kWh while coal produces almost 75% more, or roughly 2.1 pounds of CO_2 per kWh. If the marginal fuel used to generate the energy to power an EV comes from coal, the EV produces approximately 0.59 pounds of CO_2 per mile. In other words, the EV recharged with energy produced at a coal-fired generating plant would be responsible for producing as much CO_2 per mile as a gasoline-powered car getting only 33 miles per gallon.

The figures provided in Exhibit 4.1 show that the amount of energy consumed in the production of electricity from natural gas and coal are roughly comparable: 10.2 quads from coal (46.6%) and 11.7 quads from natural gas (53.3%). Using this mix of fossil-fuel sources to recharge the EV batteries results in emissions from the EV equivalent to the emissions from a gasoline-powered car getting 43 miles per gallon.

You may be wondering why I'm calculating the CO_2 emissions from an EV recharged from a grid drawing upon fossil-fuel generating resources. The reason is that virtually all EVs everywhere are being charged off a grid – a grid that's predominantly powered by fossil fuels. Since EVs are a new demand that's being added to the grid, it's appropriate to look at the marginal fuel(s) being used by utilities to supply that new demand. In nearly all cases, the marginal fuel to supply new loads is either coal or natural gas (and most likely natural gas).

Incidentally, analyses like those just presented demonstrate that a "plug-in" hybrid (recharged off the grid and/or powered by gasoline) is also not appreciably better than a high mpg gasoline-only vehicle in terms of energy consumption and CO_2 emissions.

There are a lot of factors impacting the energy consumption for EVs: the size and weight of the vehicle, the size and weight of the battery, the on-board air conditioning and heating systems, battery charging system efficiencies, driving habits, power plant efficiencies, electric transmission and distribution line losses, etc. All these factors have an impact on results like those I've just presented. I'm hoping that critics won't quibble about minor differences and at least acknowledge the basic point: EVs and **high mileage** gasoline vehicles are NOT appreciably different in terms of their environmental impact at the present time. Both types of vehicles are contributing to total fossil-fuel energy consumption and both types of vehicles are responsible for GHG emissions. When a completely decarbonized electrical grid is used to recharge EVs, then and only then, can EVs be legitimately characterized as "zero-emission" vehicles. Until that happens, EVs are typically only as "clean," or only modestly cleaner than high mileage gasoline-powered vehicles. Or, conversely, in terms of CO_2 emissions, EVs are only marginally less "polluting" than high mileage gasoline-powered vehicles.

This bit of news may be disconcerting to many EV owners who thought they were driving around carbon-free. They're not. **Their vehicles are certainly less polluting than the** average gasoline-

> *Only after the grid is predominately powered by carbon-free resources, can EVs claim to be "zero-emission vehicles."*

powered vehicle, <u>but</u> they are not zero-emission vehicles. This is true today whether the EV is charged off the grid or from a rooftop solar array.

As "early adopters" however, EV owners today can take solace from the fact that they are contributing to the long-term effort to address the problem of emissions from transportation. Specifically, by purchasing an EV, they made or are making the EV manufacturing suppliers and industry a profit-making (or future profit-making) operation. In other words, without their purchases as early adopters the EV industry would never have gotten rolling. As the electric grid becomes greener over time (i.e., emits less CO_2), all the electric vehicles on the road will also become greener.

The governor of California, Gavin Newsom, has issued an Executive Order phasing out the sale of new petroleum-powered cars and light trucks in California effective 2035. While I'm sure the governor thinks he's done something significant, I doubt that he really has.

First of all, Gavin Newsom will not be the state's governor in 2035 and will not be able to be held accountable when the governor at the time will have to decide if prohibiting the sale of gasoline powered vehicles really makes sense.

Second, this wouldn't be the first time the state has made a clean transportation pledge, only to quietly let the deadline pass without having reached the desired goal. In the mid to late 1990s, authorities in California sought to require that EVs comprise a minimum of 2% of automobile sales by the year 2000 (or thereabouts). When the magical date arrived, there weren't any commercially viable EVs even on the market. Consequently, with a lot less fanfare than was afforded the original pledge, the EV sales requirement was quietly dropped, and everybody moved on.

Third, as previously discussed, EVs are not zero-emissions vehicles unless the grid recharging them is carbon-free. If that milestone is not reached, even if Governor Newsom's executive order is still in effect, the prohibition on the sale of gasoline-powered vehicles sales will not eliminate CO_2 emissions from new cars.

Fourth, and finally, the Executive Order relates to new car sales in California. Gasoline-powered vehicles could, presumably, still be purchased outside the state and driven into California. If people are willing to drive across state lines to buy cigarettes in neighboring states with lower sin taxes, surely they'll be willing to drive out of state to buy the kind of vehicle they really want. Even with laws in place to discourage such behavior, never underestimate the ingenuity of people determined to get around a restriction that's intended to prevent them from doing what they want to do.

Before moving on, I'd like to return to this issue of announcing dramatic goals to accomplish something by a specified date in the distant future. The executive order issued by Governor Newsom or the "net zero" press releases by corporations like Amazon are examples of such announcements. Declaring that something will be accomplished at some date in the distant future is entirely different than committing to making annual progress to the ultimate goal, i.e., something that can be tracked in real time. Governor Newsom's executive order to end gasoline-powered vehicle sales in the state by 2035 would have been entirely different had the order required annual progress to the goal of zero petroleum-powered vehicle sales by the 2035 date (i.e., EVs constituting 7% of all vehicle sales in 2023, 14% in 2024, 21% in 2025, 29% in 2026 and so forth to 100% in 2035). The point is: **long-term goals without at least annual milestones are meaningless.** Such announcements are designed to make people feel good, not actually accomplish anything. When you hear such pronouncements, alarm bells should go off in your head.

EVs – A Fundamentally New Electrical Load

As previously discussed, decarbonizing even just the existing electrical grid in the next few decades will be a major challenge, so building the additional carbon-free electric generating capacity to recharge all the new EVs, **a fundamentally new electric demand**, will be nearly impossible to have in place in three decades.

Exhibit 4.1 showed that the energy content of the petroleum products consumed to move us and our goods from point A to point B is roughly 28.2 Quads. Over 80% of the source energy released by burning gasoline in internal combustion engines is rejected as heat energy. Only about 6 Quads of that energy is actually used to move our cars and trucks, etc. down the

> *Supplying the electrical energy to recharge the batteries of an all-electric mobility fleet will require carbon-free resource additions roughly equivalent to all of our current fossil fuel generating capacity.*

highway. Decarbonizing the transportation sector would require deploying enough new, carbon-free generating resources to produce 6 Quads of electric energy. Just to provide some context, 6 Quads of useful electric energy is roughly equal to all the (retail) electrical energy generated in the U.S. from coal AND natural gas combined. Using the same kinds of calculations as in the previous chapter, delivering this much energy (6 Quads) from rooftop mounted solar panels would require the installation of **3.8 billion new solar panels OR 236,000 new medium-size wind turbines OR 216 new large nuclear power plants**. (Note: this is in addition to the solar panels or wind turbines to "green" the existing electrical grid). In addition to these generating resources, there would again have to be comparable investments in energy storage capacity and transmission lines.

In addition to the fact that EVs are NOT really zero-emission vehicles and that we are not on track to deploy the additional carbon-free resources to recharge them, there's another piece of "bad news" for environmentally minded EV owners, and indeed, for all of us.

Visualize a Tesla Towing an F-150 Pickup Truck

Not only are EVs NOT "zero-emissions" vehicles, they are also responsible for greater emissions from other vehicles by way of programs intended to reward some kinds of businesses for reducing emissions and penalize other kinds of businesses that continue to emit CO_2. The programs are known as "cap and trade" or regulatory "carbon credits." Under these programs, the government sets a limit (a cap) on emissions from a particular enterprise or region. These caps are supposed to decrease over time. In practice, however, they're often allowed to remain the same if the local economic consequences of a reduction would be too great. The "trade" part of the program allows enterprises whose emissions are below some target value to receive **credits**, which they can then trade (i.e., **sell)** to enterprises that are unable or unwilling to reduce emissions below their allowable limits. Regulatory carbon credits work similarly. The credits allow enterprises that are clean or have made investments to reduce emissions to sell their credits to high polluters for a profit. The enterprises earning the credits often advertise their environmental credentials to gain public approval and increase their market share. These same enterprises typically don't advertise (or may even try to conceal) details about the revenues they derive from the sale of their carbon credits. Their good environmental citizenship is actually just a mechanism to increase their bottom line by selling "credits" that allow other enterprises to continue selling their high polluting products into the marketplace. The programs just become a mechanism to transfer funds from the deep pockets of one industry to the deep pockets of another without markedly improving the environment.

One of the companies enjoying the proceeds of regulatory carbon credits is Tesla. Tesla can sell carbon credits for its EVs to any high polluter. One of those potential customers is the Ford Motor Company. In order to meet fleet-wide mileage standards or emissions reduction requirements and avoid hefty fines, Ford can purchase carbon credits from a company like Tesla in order to allow it to keep selling its higher polluting and higher profit gasoline-burning pickup trucks in the lucrative California market. The revenues Tesla received by selling those carbon credits were reportedly a significant fraction of its revenues in the early quarters of the company's existence when, based on car sales alone, Tesla would have reported losses for multiple quarters.

Tesla receiving carbon credits for selling EVs strikes me as a gross distortion of regulatory "carbon credit" programs and the marketplace. The programs were supposed to

> *Giving Tesla (or any EV manufacturer) carbon credits for building EVs would be like paying McDonalds to sell hamburgers.*

promote energy saving and emissions-reducing improvements to business operations. Tesla is in business to build and sell EVs. That's what they do. Allowing them to claim carbon credits for building EVs would be like paying McDonald's to sell hamburgers. **If anybody should get carbon credits related to EVs, it shouldn't be the**

manufacturer. It should be the customer who operates the EV because it's the EV owner who paid a premium for the EV in the first place and it's the EV owner who's displacing CO_2 emissions by driving the EV rather than driving a gasoline-powered vehicle – not the manufacturer of the EV. We should only be providing tax or environmental credits if failing to do so would prevent a beneficial outcome from being realized. If we stopped giving "carbon credits" to EV car manufacturing companies, would they stop building EVs? I don't think so.

I don't know why owners of Teslas (and other EVs or hybrids) haven't banded together to DEMAND that THEY be granted the carbon credits rather than allowing them to flow to the manufacturers. (Call to action?)

I understand the concept behind cap-and-trade and regulatory carbon credit programs to provide additional revenues to companies that are helping to reduce CO_2 emissions, but these programs are difficult to administer and monitor. Companies or individuals, e.g., farmers, can manipulate their own data or processes to overstate emissions reductions to earn more credits. (Of course, this kind of creative accounting is not limited to the carbon credits market). Investigative journalists have also identified cases that raise questions about the longevity of claimed carbon credits in the real world.

A final concern about carbon credits programs is that corporate entities are generally more skilled at inserting loopholes into and exploiting weaknesses of legislative and regulatory programs than legislators and regulators are at crafting those programs. Teams of paid lobbyists shape these governmental programs so that they are least harmful to their mother companies. Teams of corporate lawyers pick through regulations looking for loopholes more exhaustively than a flock of vultures devouring a carcass. Simpler and more straightforward programs are better because they are less subject to manipulation. Sales taxes are a perfect example: you buy something, the store collects the sales tax, the revenues are transferred to the government, end of story.

There's A Better Way

A more straightforward method of reducing CO_2 emissions would be to implement a carbon tax. I'll discuss this approach in a later chapter, but for now, suffice it to say that a carbon tax is far simpler, more effective, faster acting and less prone to manipulation than a cap-and-trade or regulatory carbon credits programs. The most significant drawback of a carbon tax, however, is that it's visible. This visibility is one of the reasons it's unlikely to ever be adopted. Nevertheless, if you need results fast, which we do, a carbon tax is the way to do it. To avoid calling it a tax, some people refer to it as "carbon pricing;" it's a tax.

When and Where EVs Get Recharged is Important

The personal electric mobility option (i.e., excluding electric mass transit - subways, light rail, people movers, electrified heavy rail trains) is, as previously noted, an (almost) entirely new demand for electric energy, completely separate from the existing electric utility grid. To power this new demand for clean electric energy in the next few decades, we're stuck with the same technology options available to us to repower the existing electric utility grid: specifically, solar and wind power. But the electrical demand associated with transportation introduces some new wrinkles.

EV owners will want the ability to recharge their EVs at home, but most vehicles, including EVs, leave their home base on most working days. This means that most EVs being recharged at home will have to be recharged from energy sources operated in the early morning hours, late afternoon hours and/or overnight. Wind sources may or may not be available at the time this energy is needed, but these hours are most definitely not prime solar production hours. Consequently, for the near future, the average EV is most likely going to be recharged from the grid served by generating stations burning fossil fuels.

The alternative to recharging EVs from the grid at night is to recharge them from the owner's personal storage battery (like the kind Tesla and others are now marketing for home energy storage). In other words, to bypass recharging their EV from the grid at night, the EV owner would be plugging his/her EV into the energy storage battery mounted on the wall inside their garage. Energy would be flowing from the home storage battery to the EV storage battery. In this example, the EV owner would have to purchase an additional home energy storage battery (for EV charging) and additional solar panels to charge these storage batteries during the day. This scenario raises some other issues that are exactly comparable to the electric grid issues.

How many storage batteries dedicated to recharging an EV will an EV owner need? Will it be sufficient to add just enough storage capacity to recharge the EV battery from "empty" to "full" one time? Or will it be necessary to buy two, three or more sets of batteries to ensure that the EV owner will always have enough stored energy to recharge their EV even if it's rainy or overcast for multiple days in a row? And how many extra solar panels will need to be installed in order to recharge the dedicated EV batteries and their backup batteries and over what period of time? If the EV owner installed three sets of EV recharging batteries, will that owner want to be able to recharge that entire capacity in one day? In addition, an EV battery pack has thermal management requirements. In warm areas, battery packs may need to be cooled to ensure that they don't overheat. In cold climates, that same battery may need to be heated so that the battery is capable of performing properly.

The workday work-around for some EV owners wanting to recharge their vehicles during daylight hours is to recharge them at work. But that would require employers to install and maintain charging stations connected to the grid or install solar arrays dedicated to this service. Many companies have installed EV recharging stations for their employees, but typically they've installed only a relatively few recharging stations, matching the current need. What happens when EVs make up 20% or eventually even 50% of all employee vehicles? In any case, unless the EV owner works for a Silicon Valley tech firm or other "generous" company, there's going to be a charge for this benefit. At least it could allow the EV owner to forgo having to invest in a dedicated, home-based energy storage system exclusively for their EV. Maybe you've heard the expression: "There's no free lunch." (And even when you think it's free, it's not really free.)

When Waste Heat Is Not (Totally) Wasted

In order to provide adequate power for acceleration, internal-combustion-engine-powered vehicles generally have a lot more power than is needed to push the vehicle down the highway at a constant speed. Hybrid vehicles achieve much of their efficiency advantage by using smaller internal combustion engines running at a constant speed to recharge the on-board battery when excess energy is available and combine forces with the battery to provide the extra power needed for acceleration.

In addition to providing propulsive power to move the vehicle down the road, internal combustion engines also generate a lot of heat, the vast majority of which is discharged directly to the atmosphere without having done any useful work. In our petroleum-powered cars, some of the excess energy released in the combustion process can be diverted to the passenger cabin to provide heating during cold weather. Similarly, the excess engine power from an internal combustion engine powers the vehicle's air conditioning system when it's hot. Neither of these two functions, heating or cooling, can be achieved in an EV without extracting energy from the battery. In addition to these cabin-comfort features, energy may also need to be extracted from the on-board battery to heat the vehicle's battery in cold weather locations or cool the battery in warm weather to prevent it from overheating. All these auxiliary loads that are either non-existent or not generally important in an internal-combustion-engine-powered vehicle become much more important in an EV because the energy they pull from the battery shortens the EV's effective range, degrades the EV's environmental benefits and increases the number and/or depth of recharging cycles.

The vast majority of EVs will be recharged from the electric grid and will, accordingly, carry the burden of the CO_2 emissions characteristic of the grid resources. However, even EV owners who "go the extra mile" to install a home storage battery and extra solar panels to recharge that battery will most likely still have to maintain a grid connection and pay a premium to do so.

It's also worth remembering that batteries typically have an upper limit to the number of charging/discharging cycles. Modern EV manufacturers warrant their batteries for several hundred thousand miles and claim that the original EV owner will probably never have to replace the battery for as long as they own the vehicle. The efficacy of those claims for the original owner and any subsequent owners will have to be tested over time. In any case, the cost of a wholesale battery replacement may depress the resale value of any EV since the new owner won't know how well or how poorly the original owner cared for the battery.

By pointing out some of these issues with EV, you may conclude that I'm not a fan. **That would be wrong**. EVs produce (at the power plant) only about half the amount of CO_2 produced by the average gasoline-powered vehicle on U.S. roads today and that's a good thing. I also worked in the Electric Transportation Research program at Southern California Edison. As part of that program, I drove an EV van (albeit a far cry from any modern EV) to and from work collecting data for several months, **thirty years ago**! I am well aware of both the pros and cons of EVs and fully expect that my next car will be an EV – and not only because of reduced emissions of all kinds, but also the potential for significantly lower maintenance expenses and headaches. In the same period while I was part of the company's electric transportation research program, I also created and then led an Advanced Electric Transportation Research Program. Our mission was to evaluate the potential for magnetic levitation passenger trains, bullet trains, electrified heavy rail systems, electrified bus lines and others. At the time, improving the air quality in Southern California was the primary objective for these programs, not reducing CO_2 emissions.

The introduction of massive numbers of EVs (cars, light trucks, buses) raises a number of policy issues. Currently gasoline taxes are used (in part) to maintain roads and highways. As EVs become a larger and larger share of the vehicles on the road, which they will, it will probably be necessary to extract some form of a road fee from EV owners as a substitute for revenues previously derived from gasoline taxes. Indeed, the entire concept of gasoline taxes may need to be revisited. In Europe, taxes nearly double the cost of gasoline compared to what is here in the United States. Is it any wonder European and Japanese car makers led the way to smaller, more fuel-efficient vehicles?

Beyond Personal Mobility Options

More mass transit systems will be built and expanded in coming years and existing bus networks will be electrified. Both of these trends will drive up electricity demand but will help reduce carbon emissions after the grid powering them is decarbonized.

But the movement of people is only one aspect of electrified transportation. One of the projects I proposed as part of the Advanced Electric Transportation Research Program was electrification of the freight rail lines from the ports of Long Beach and Los Angeles to some inland location such as Barstow. We were never able to get the heavy rail electrification project green-lighted, but the need and the opportunity are still there.

Every major rail line in the United States should be electrified. This kind of a project is not as flashy as solar installations but could yield significant and sustained environmental benefits. Electric locomotives would replace their diesel cousins. Like EVs, however, this load would be a new electric demand requiring the same kind of investment in green generating capacity, energy storage and transmission lines to make it sustainable in the long-term. And, like EVs, fully electrifying the railroads would not yield significant CO_2 emissions reductions until the sources of the electric energy to power those trains were also carbon-free.

While the majority of CO_2 emissions in the transportation sector come from cars and trucks, there are also significant CO_2 emissions from ships, planes, and heavy equipment. As previously noted, trying to power the entire transportation sector with agriculturally derived fuels wouldn't be possible or make sense. However, devoting a portion of agricultural output to the production of transportation fuels for aviation and other commercial applications like powering farm equipment (if nothing else exists) may make sense. In addition, hydrogen and ammonia are being tested for use as a transportation fuel for ships. "Chemically synthesized," liquid, mobile fuel products are also feasible if, as previously noted, there were some high temperature, carbon-free process heat source available (such as a nuclear reactor).

Permission To Come Aboard, Captain

In addition to providing process heat for liquid fuel production, there's another way nuclear power could fill a niche in the marine transportation sector. Nuclear technology for civilian marine applications is a distant dream, but not for lack of technological development. Large numbers of nuclear-powered ships, mostly military, are currently operating around the globe, but decades ago the civilian, U.S.-built, nuclear-powered NS Savannah (in service in 1962) and the German-built nuclear-powered Otto Hahn (1964) cargo ships operated successfully as technology demonstrations.

Just as advanced nuclear concepts to generate electricity will receive more attention in coming years, nuclear powered ships may also become part of future civilian maritime fleets. But, as with land-based nuclear reactors, cost and public acceptance issues will prevent the re-introduction of maritime nuclear applications in time to help combat global warming. A fleet of nuclear-powered commercial transport vessels could be plying the seas by 2100, but that will already be 50 years too late to avert the onset of the worst global warming impacts.

[*Input] Individuals familiar with non-CO_2 producing transportation options are requested to submit articles for consideration for inclusion in the 2nd edition of this book. [*Input - End]

A Mini-Recap

Eliminating CO_2 emissions associated with the transportation sector is absolutely essential to "beat" global warming and the key to making that happen in a timely fashion requires two critical developments:

1. Electrifying (almost) everything that moves, and
2. Producing the energy to power those mobility options (or used to manufacture the synthetic, liquid, mobile fuels that go into them) from exclusively carbon-free resources.

While the year-over-year percentage growth in EV sales has been impressive, the total annual EV sales are still minuscule and concentrated in unique markets like California, which, in the early days of production, reportedly accounted for 90% of all domestic Tesla sales. EVs of all shapes and sizes from a whole host of manufacturers have become a larger and larger fraction of the worldwide market in a very short period of time. Electrification of the entire world's fleet of cars and light trucks will be very gradual. A sizable percentage of the internal-combustion-engine-powered vehicles rolling off assembly lines around the world today are likely to still be on the road spewing out CO_2 for the next 10 or even 20 years. Even hybrids, which generally boast high mpg ratings (i.e., good fuel economy), still burn fossil fuels. The majority of the vehicles on the world's roads and highways in 2050 will still run on petroleum.

On the bright side, the U.S. and the world are moving credibly toward a personal mobility sector dominated by electric vehicles. In the long-term, they will help reduce CO_2 emissions, but their full carbon-free potential will only be realized if those vehicles are recharged from carbon-free resources. Until then, EVs will be only marginally better than gasoline-powered vehicles in terms of CO_2 emissions.

At present, neither of the two requirements to substantially reduce GHG emissions from transportation sources are being met: electric vehicles are not penetrating the market quickly enough and the grid is not being "decarbonized" rapidly enough. (In contrast, natural gas and even coal-fired electric generation is still expanding worldwide). Some investments are being made in transit systems, but too few and too slowly and virtually nothing has been done to electrify the movement of goods by train in the U.S. Regrettably, the ever-increasing numbers of cars and light trucks on the road will partially or maybe completely offset any emissions reduction achieved through EVs, hybrids and higher mpg vehicles.

In short, we're not winning the emissions war in the transportation sector.

Action Items

As relates to the mobility issues discussed in this chapter, here are a few of the projects and policy initiatives that could help address transportation-related climate stresses.

1. Increase the per gallon gasoline taxes by $0.30/year for at least the next 10 years ($3.00/gallon over 10 years). Use the additional revenues to accelerate the decarbonization of the utility grid and advance other transportation initiatives to reduce CO_2 emissions.

2. Increase Corporate Average Fuel Efficiency (CAFE) standards for cars and light trucks. [Three points: 1) Raising gasoline taxes by $0.30 per gallon per year each year for the next decade as suggested above would eliminate the need for the government to impose higher CAFE standards. Car buyers themselves would demand high efficiency vehicles; 2) I never said these recommendations would be popular. I'm simply making the point that if we want to slow global warming, we're going to have to do some things that will not be popular and will drastically alter our current lifestyles. 3) Most of the economic changes to combat global warming will be regressive (i.e., they will be more costly and more impactful to lower income individuals) unless special efforts are made to ensure an equitable distribution of "pain" for one and all. The rich might complain, but they'll be able to pay; the rest of us will struggle.]

3. End ethanol production within five years. The transition period would give investors a chance to recover some of their investment. In the not-too-distant future, food shortages will increase. All productive farmland needs to be allocated to crop production for food, not mobility.

4. Electrify all mainline and coast-to-coast train routes within 10 years. Migrate all long-haul truck traffic to the electrified railroads.

5. Increase mass transit capacity and efficiency. Make mass transit free to riders.

6. Require airports in cities greater than 500,000 to have mass transit links to their city centers.

7. Revise zoning laws to encourage high-density, walkable living spaces.

8. Increase EV licensing fees to be equivalent to gasoline taxes for road maintenance and to cover the extra expenses of special fire-fighting requirements for EV fires. This won't increase EV sales, but it will ensure that EV owners shoulder their responsibilities for road maintenance and safety.

Transportation

12 RESIDENTIAL, COMMERCIAL AND INDUSTRIAL INTERESTS

"What's good for business ~~is good for America.~~"
[. . . may not be good for the planet.]

A rewriting of a statement by Charles E. Wilson (President Eisenhower's Secretary of Defense and a former CEO of GM)

CO_2 emissions don't come only from electric generating stations burning fossil fuels and petroleum-powered transportation sources. Exhibit 6.1 categorized the sources of greenhouse gas (GHG) emissions and the energy flow graphic, Exhibit 4.1, illustrated where that energy comes from and goes to. The EPA estimates that 22% of U.S. CO_2 emissions come from industrial processes, so "beating" climate change not only means attacking emissions from electric generation and transportation, it also means attacking emissions from residential, commercial, and industrial processes that use natural gas, petroleum, and coal directly.

Exhibit 4.1 showed that the U.S. residential sector in 2019 consumed 7.0 Quads, the commercial sector consumed 4.8 Quads and the industrial sector consumed 23.1 Quads of **non-electrical** energy annually. Combined, that's almost 35 Quads of non-electrical energy content flowing to these three sectors. Not all of the carbon-based materials extracted from coal mines and oil wells that are consumed in the industrial sector are burned. Some fraction of this material is consumed as a raw material (a feedstock) in, for example, the manufacturing of plastics. If this carbon-based material is not burned, it doesn't contribute to CO_2 emissions (at least not directly). If 25% of the fossil fuel material consumed in the industrial sector is used as feedstock chemicals (i.e., not burned) and all of the energy produced from biomass is subtracted (because any carbon content in it came from the atmosphere) that leaves over 27 Quads of carbon-based energy consumed in these three sectors in the U.S. that has to be replaced with an electric alternative. Assuming fossil fuels burned directly in industrial, commercial, and residential settings is only 70% efficient and that electrical energy replacing the fossil fuel is 100% efficient, carbon-free resources would have to provide 19.1 Quads of energy to displace fossil fuels consumed directly (burned) in these sectors.

Using the same kinds of calculations as in the previous two chapters, delivering 19.1 Quads of energy from rooftop mounted solar panels would require the installation of **12.2 billion new solar panels, or 750,000 new medium-size wind turbines, or almost 700 new large nuclear power plants.** (Note: this is in addition to the solar panels and wind turbines to green the existing electrical grid and displace petroleum from the transportation sector). In addition to these generating resources, there would again have to be comparable investments in energy storage capacity and transmission lines.

The carbon-based energy consumed directly in the residential, commercial, and industrial sectors is burned in boilers, furnaces, ovens, kilns, etc. This energy is used in drying processes, heating materials and working fluids, steam production, space heating, food preparation and many, many other uses. Electrical resistance heating can be used in most of these applications but consumes between two and three times as much source energy to perform the same functions provided by the direct combustion of fossil fuels. Advances in electric heat pumps make them a possible alternative for space heating and water heating though their efficiency degrades in extremely cold areas. Still, the direct burning of natural gas is anywhere from 40% to 60% more efficient than utilizing advanced heat pumps. The lower fuel efficiency for the electric alternatives in all these cases translates into greater CO_2 emissions. In other words, if I replaced my gas-fired water heater with model using electric resistance heating, I would more than double my CO2 emissions; if instead I used an advanced heat pump my emissions would increase by roughly 50%.

In any case, eliminating CO_2 emissions from the combustion of fossil fuels in the residential, commercial, and industrial sectors takes this discussion directly back to the electrical grid, renewable energy resources and energy storage. (Sound familiar?). All of these topics were discussed in Chapters 6, 7, 8 and 9.

These numbers can vary from appliance to appliance or from industrial process to industrial process, but the message is the same: if the electrical energy you're consuming to displace the direct combustion of natural gas is not carbon-free, your CO_2 emissions will go up, at least modestly and often dramatically.

This leaves us with a "Catch-22." If you want to reduce carbon emissions, you have to electrify everything. But, if the grid supplying that electrical energy is burning any fossil fuel, you're almost certainly increasing your carbon footprint if you electrify an appliance or a process previously provided by burning natural gas directly.

Another consideration for commercial and industrial processes is that they operate most efficiently when they run (virtually) continuously. If a process is converted from the direct combustion of a fossil fuel (e.g., natural gas) to an electrical process, the reliability of the electric grid is no longer just desirable, it's essential. If the reliability of the electrical grid drops, it may be necessary for the enterprise to install its own energy storage capacity to sustain the industrial process when there's a grid outage. In short, satisfying the energy demand requirements of the commercial/industrial sectors may be one of the most critical hurdles in displacing fossil fuels in these end uses.

The production of chemicals (such as chemical fertilizers), steel and cement are all major sources of industrial CO_2 emissions.

As Hard as a Rock

One of the least recognized industrial sources of CO_2 emissions comes from manufacturing cement. The BBC reports that the manufacturing of cement makes up 8% of all worldwide CO_2 emissions. (They report that if cement manufacturing were a country, it would be the third largest emitter of CO_2 behind China and the U.S.)

Concrete (sand, water, aggregate and cement) is the most widely used construction material in the world. (Cement is the ingredient that holds concrete together). Most cement is manufactured in industrial kilns into which pulverized limestone and clay (and other additives) are fed in at one end, heated to temperatures in excess of 1400° C, and expelled at the other end as "clinkers." Cement is produced by grinding up these clinkers. The chemical reaction that occurs in the kiln is called calcination and is a source of the CO_2:

$$CaCO_3 \text{ (limestone)} + \text{Heat} \longrightarrow CaO \text{ (ingredient in cement)} + CO_2$$

Because the clinkers are the product of a chemical reaction, it's not possible to eliminate this portion of CO_2 emissions from the process.

This chemical reaction is responsible for roughly half of the CO_2 emissions from the production of cement. Most of the rest comes from the combustion of some fuel (e.g., natural gas) to provide the extremely high temperatures required to drive the chemical reaction.

There are some promising technologies on the horizon to reduce CO_2 emissions from the cement production process. These include the use of additives in the calcination kiln to allow the reaction to proceed at a lower temperature so that less fuel has to be burned to provide the heat to drive the chemical reaction. Another is to use alternative fuels such as hydrogen as the heat source, as long as the hydrogen is not produced from natural gas. And yet another is to alter the clinker-to-cement process. These technologies are vitally important because the 8% contribution to global CO_2 emissions today from cement production is predicted to increase dramatically as worldwide demand for concrete and cement grows. The website carbonbrief.org reports that "the floor area of the world's buildings is projected to double in the next 40 years." In other words, even with improvements in the cement production process to reduce the number of tons of CO_2 per ton of cement produced, the total CO_2 emissions from this sector will increase markedly, not DECREASE, as total cement production worldwide increases.

As Strong as Steel

Another industrial process responsible for enormous CO_2 emissions is steel production. For each ton of steel produced, modern steel furnaces release almost two tons of CO_2. Steelmaking is responsible for 7% to 10% of all CO_2 emissions worldwide, i.e., comparable to the emissions from cement production. In steelmaking, hydrogen could replace coal and coke and has the potential to eliminate virtually all CO_2 releases from the process, but again, only if the hydrogen is "green" hydrogen. (Recall that 95% of the world's current production of hydrogen comes from natural gas which generates massive CO_2 emissions.)

Water, Water (Needed Almost) Everywhere

The production and distribution of freshwater is another area that has significant energy and emissions implications. In future years, freshwater supplies are almost certainly going to become much more critical than they are today. The over-pumping of aquifers in the U.S.

and around the world that's already taken place is forcing farmers and cities to drill deeper and deeper wells.

In dry or desert regions, localities have turned to reverse osmosis (RO) and multi-stage flash distillation to generate freshwater from brackish or sea water. Both processes use a lot of energy. RO uses energy to pump salt water at high pressure across specialized membranes to produce freshwater. Flash distillation plants use energy to boil sea water, flash it to steam and then collect the freshwater condensate. Most distillation processes use heat from the combustion of fossil fuels or "waste heat" from electricity production, also from fossil fuels, in what's called a bottoming cycle. In either case, there are CO_2 emissions associated with the production of freshwater by these technologies. In Saudi Arabia, cheap fossil fuels make such processes economic. In the future, after CO_2 emissions are taken into account, arid regions of the world will have to find other means to acquire or "manufacture" the freshwater their citizens need.

If we successfully stop burning fossil fuels to produce electricity, the low-grade heat energy used in bottoming cycles, such as the flash distillation process to produce freshwater, will no longer be available. This lower temperature energy will also not be available for other applications such as district heating, process steam production, hot water heating, warming greenhouses, etc. In a world dependent upon solar panels and wind turbines, none of these low-grade energy bottoming cycles will be viable. The investments in bottoming cycle plants will be stranded and more importantly, these facilities will no longer be able to produce the products or services that people may have come to depend upon.

Moving Water from A to B

Aside from thermal desalination and reverse osmosis, transporting water from regions of abundance to regions in need is also an option. That, of course, is what allowed Southern California to flourish after the construction of the California Aqueduct in the early 1960s. In the future, the mass movement of freshwater from region-to-region will almost certainly increase. That water will have an unseen energy component associated with running the massive, industrial-scale electric pumps to move it from place to place. As with EVs, this new electrical load will require even more green electric generation capacity.

A "Diet" Rich in CH_4 (Methane)

Another commercial/industrial enterprise that generates significant quantities of CO_2 and methane is the agricultural sector. Producing, processing, packaging, and transporting our food accounts for 10% to 20% of the nation's total greenhouse gas emissions. Within the agricultural sector, meat protein and dairy products are responsible for a particularly large fraction of those emissions. This is because more than 50% of U.S. grain production is used to feed livestock. For the world as a whole, 40% of total grain production to feed farm animals. Growing and harvesting that grain produces CO_2, but in addition, dairy cows and beef cows have unique digestive systems that produce methane gas as a by-product of converting grain, grass, and hay into the energy they need to live and grow.

Experts estimate that it takes 6 pounds of grain and over 10,000 gallons of water to produce a single pound of high-quality meat protein. These livestock animals are also responsible for 22 pounds of CO_2-equivalent GHG emissions, mostly methane, **per pound of beef**. This is why climate change activists encourage meat-eaters to adopt a meat-free day in their weekly diet or, better yet, become vegetarians.

When we buy gasoline or pay the electric bill, we know we're paying for energy. But just as in the case of water and food, it's easy to overlook the "energy content" in many commodities such as clothes, consumer items, manufactured goods, etc., but there's an energy component in just about everything we consume. As energy becomes more expensive, the cost of literally everything will go up along with it. As the cost of goods and services increase compared to what they are now, people will have less disposable income and purchasing choices will gradually change. Measures to combat global warming will have ripple effects throughout the economy.

[*Input] – Individuals who are able to add additional insights in regard to the challenges of reducing CO_2 emissions from agricultural, commercial and industrial enterprises are invited to submit them via the BlueOasisNoMore.com website. [Input – End]

Delivered Directly to Your Door

One of the most dramatic changes in the commercial retail market, especially in the last decade, has been the astounding rise in online shopping. As the transition from brick-and-mortar stores to online shopping has progressed, there's been an even more astounding rise in overnight delivery services - Amazon, FedEx, UPS, USPS, etc.

The logistical sophistication, economies of scale and convenience these companies have achieved are beyond miraculous.

Exhibit 12.1   Commercial Waste

If, however, we're trying to drive down CO_2 emissions from all sources to save the planet, I'm fairly sure this obsession with next-day delivery of the products we order is dragging us in the wrong direction. I have to believe that the carbon footprint of getting a pair of scissors, a printer toner cartridge, a ream of paper, and a box of fasteners into the hands of a customer in one visit to an old-fashioned, brick-and-mortar office supplies store is a lot lower than having each of these items delivered to somebody's front door, separately, in their own cardboard box filled with crumpled brown paper or air packets. (And remember, even if your package is delivered by an electric delivery van, its emissions are not zero.)

[*Input] – Overnight Delivery Energy Analysis – If anyone's done a comprehensive, comparative analysis of the energy and material resources consumed in online shopping with home delivery vs. an in-store product acquisition process, please submit it for inclusion in the 2nd edition. [*Input – End]

The commercial and industrial uses of fossil fuels don't get a lot of play in the press, certainly not as much as wind turbines, solar panels and EVs. After all, when was the last time you read about industrial gas ovens being converted to electric ovens? Nevertheless, if we're serious about "beating" global warming, we will need to displace fossil fuels from industrial and commercial processes, and we'll have to do a lot more than just meatless-Mondays to curtail agricultural emissions.

The direct combustion of fossil fuels in residential, commercial, and industrial processes will be a particularly difficult area of fossil-fuel consumption to eliminate. Why would an industry abandon fossil fuels that are an integral part of well-honed processes and most often produce less CO_2 than their electric counterpart derived from fossil fuels? Fossil fuel is cheap, well understood and requires no additional capital investment. If there are no business disincentives for burning fossil fuels or doing something differently than the way it's been done for the last 50 or 100 years, nothing's going to change.

Action Items

To reduce CO_2 emissions in the residential, commercial, and industrial sectors, here are a few of the projects and policy initiatives that could help:

1. Establish or increase water consumption taxes. Use the revenues to build water storage/delivery infrastructure, provide subsidies for water conservation measures undertaken by consumers and to build flood control infrastructure on an expedited basis.
2. Establish standards to dramatically reduce packaging waste. Establish product repairability standards. Require all durable products to be 100% recyclable.
3. Require landscaping and agricultural production to conform to "regionally appropriate" standards, i.e., no high-water-consumption crops in drought-prone regions.
4. Require product delivery services (Amazon, FedEx, etc.) to coordinate with one another to minimize local traffic and fuel consumption. Encourage these services to combine forces to make no more than two deliveries per household per week.
5. Stop deforestation and the destruction of wetlands for commercial and agricultural development.
6. Increase residential and commercial building insulation standards by 50%.

13 ECONOMICS

"Money, money, money, money, money, money,
Money makes the world go around,
the world go around,
the world go around."

Cabaret
Song lyrics by Fred Ebb, 1966

Like many good songs, these lyrics communicate a fundamental truth. This book is about climate change, but aside from the principles of chemistry and physics, this book is also just as much about economics – and by that, I mean: costs (monetary and otherwise), benefits (monetary and otherwise), relative costs, innovation, policies, incentives, etc. All of the choices we make as consumers, all of the policies we pursue as a country, all of the enterprises in which we engage -- all of them have an economic dimension. In this book, global warming just happens to be the issue around which all of these economic considerations revolve.

The Die is Cast

Mankind has never been content with the status quo. We've always been on a quest for something different, something better. Economist Joseph Schumpeter coined the term "creative destruction" to capture the dynamic process of innovation, progress, and all kinds of changes in the marketplace of products and ideas. The quest for new knowledge accelerated when science, scientific principles and experimentation took hold leading to the industrial age. Powerful machines began replacing muscle power, eased human drudgery, increased productivity, increased prosperity and generally improved the quality of life for people everywhere. And of course, those machines were powered by fossil fuels.

At the dawn of the industrial age, before fossil fuels, there were few power-producing options. There were water wheels, simple windmills, and animal power (oxen, horses, etc. and of course, us). There were no solar cells and no electric-generating wind turbines, but **everything changed** when we began to exploit fossil fuels.

Ingenious inventors developed, and, over time, perfected, marvelous machines (first steam engines, then internal combustion engines and ultimately steam turbines and jet engines) to convert the heat energy derived from burning fossil fuels into rotational kinetic energy, i.e., energy to turn wheels and drive shafts – energy to do useful work that humans then didn't have to do.

At the dawn of the industrial age, societies unconsciously did a cost/benefit analysis for using the newly discovered fossil fuels. The benefits of using fossil fuels were readily apparent. The costs were not. The physics and chemistry of global warming were not widely known or understood at the time, so the future impacts of CO_2 emissions were simply left out of the analysis. However, even if the future impacts of CO_2 emissions had been fully understood at the time, it wouldn't have made any difference. Voluntarily choosing to forgo the benefits of fossil fuels would never have been a serious consideration. The benefits of extracting energy from fossil fuels to perform useful work were just too enticing.

Once mankind started down the fossil-fuel superhighway, the die was cast. We can't look back now from the early decades of the 21st century and curse the path that our predecessors took. We would have

> *Once we started down the fossil-fuel superhighway, there were no off-ramps. We became "addicted" and now that we understand how harmful our addiction is, (like all additions) quitting fossil fuels is not going to be pretty.*

done exactly the same. In any case, we, today, have to play the hand we've been dealt. But, while we can't in good faith curse the path our ancestors chose, our descendants may very legitimately curse our failure to act once we understood what was at stake.

Pay Me Now OR Pay Me Later

I've been characterizing the transition from fossil fuels to renewable resources as expensive, a claim that's at odds with what a number of global warming/climate change activists are expressing.

What I hope to do in the rest of this chapter is to convince you that energy prices in a decarbonized world will go up and explain why they have to. I'm not raising this issue as a reason or an excuse for not doing it! On the contrary, regardless of the expense, we need to make this transition as quickly and as completely as possible. It's the only way we'll be able to delay and lessen the worst consequences of global warming that future generations will have to deal with. As they say in that oil filter commercial: "Pay me now or pay me later." (But you WILL pay me).

I'm raising this issue of the cost of abandoning fossil fuels because promises of a cheap and easy transition that are ultimately unfulfilled could doom the transition to failure. It's better for the public to understand what's coming than to be surprised, resentful or even hostile as a more expensive energy future unfolds.

> *Unkept promises of a cheap and easy transition away from fossil fuels could doom the transition to failure. It's better for the public to understand what's coming than be surprised, resentful and even hostile when a more-expensive future unfolds.*

The Cost of Going Carbon-Free

Others have made estimates of the cost of getting to a carbon-free future. Those estimates range anywhere from tens of trillions of dollars to hundreds of trillions of dollars over X number of years, but honestly, when anyone starts talking about trillions of dollars, my eyes start to glaze over. I have no way of really relating to those kinds of numbers. What I can relate to is this: "How much of my monthly budget goes to paying my energy bills? Whatever that dollar amount is, it will be at least a factor of three or four times that amount in the future. And that's just for the direct consumption of energy. We'll also be paying three or four times as much for the energy content in everything you consume – food, clothes, shelter, refrigerators, computers, books . . . everything.

The Big Picture

Let me begin by asking a simple question: would we be talking about making the transition from fossil fuels to exclusively carbon-free energy resources if it weren't for global warming concerns? In other words, if burning fossil fuels didn't produce CO_2, would we be thinking about deploying solar panels, wind turbines, energy storage facilities and transmission lines in mind-boggling numbers?

With the exception of isolated islands and some other remote locations, generally speaking, the answer is: "No."

The point is this: the world's economy has evolved based on readily available and comparatively inexpensive fossil fuels. The technologies to use these fuels have been refined over many decades and a massive infrastructure has been constructed to exploit them. For better or worse, the system we have today is mostly as efficient as it can be. We've settled into a "least-cost" operating point. If there were a better, cheaper way to run things, the system would have already gravitated to it or would be in the process of doing so. As innovative technologies evolve and enter the marketplace, that least-cost operating point shifts slightly. It's always shifting. If there were a "silver bullet" to solve the global warming crisis, there'd be no need for rebates, tax incentives or overly favorable solar buy-back rates to accelerate the movement toward that new least-cost operating point. It would just happen (over time) because it was truly less expensive.

But that's not what's happening now. We're **not** proposing to abandon energy-dense fossil fuels because there are less expensive carbon-free alternatives that can deliver the same level of system reliability on which modern economies depend. We're turning away from fossil fuels because continuing to burn them is killing the planet.

> *We're "forcing" ourselves to abandon fossil fuels for more expensive alternatives because burning fossil fuels is killing the planet.*

Moving off this "least-cost, fossil-fuel-based, operating point" will increase energy costs. Up until now, CO_2 emissions have been treated as having a zero cost (i.e., what economists call an "externality"). Initially, we didn't understand the cost this externality would impose on future generations, but that cost is becoming increasingly clear. Anytime an externality is reintroduced into a cost equation, expenses go up. The planet will be better off for it, but the energy we want and need will be significantly more expensive.

To illustrate the impact of introducing a previously ignored process externality, consider the case of General Electric which, over a thirty-year period, dumped more than a million pounds of PCBs into the Hudson River. GE was not alone in dumping its untreated wastes. Numerous other industrial players did the same, to the Hudson and every other major river around the globe. While disastrous for the river and everything that previously lived in it, and oh yes, the people who just happened to live downstream, GE's practice of indiscriminately dumping its waste rather than incurring the cost of properly treating its effluent in-house, boosted its bottom line. By avoiding an added cost of production, GE increased its profits. Once GE was prohibited from polluting the Hudson River, its production costs went up and it either had to raise the price of its product or accept a lower profit margin. This is also what happened at coal-burning plants that had to retrofit pollution control equipment to remove sulfur dioxide (a precursor to acid rain) and fine particulates from their smokestack emissions.

Back Up the Truck

While it's clear to me that the cost of energy in a world free of fossil fuels is going to be a lot higher than it is now, I realize that some of you may not be buying my argument, so let me try coming at this a slightly different way, i.e., by looking at what's been happening at world climate change conferences.

At those conferences, representatives of the participating nations (basically all of the nations of the world except the U.S. in the period from 2016 to 2020) were attempting to negotiate carbon emissions reduction targets.

> *When the cost of energy goes up, the price of everything that has an energy-cost-component also goes up. In other words, the price of everything goes up.*

The negotiating strategy for every nation was to minimize its own commitment to reduce carbon emissions while providing a rationale why every other nation needed to increase theirs.

The modest reduction targets set at the Paris Climate Conference were voluntary – i.e., they were a tentative first step, but at least a move in the right direction. Those reduction targets

> *If a wholly "green" future, free of fossil fuels, really is a "least cost option," why wouldn't the world's representatives be falling all over each other to commit to massive emissions reductions?*

were the diplomatic climate equivalent of "baby steps." Everyone recognized that these goals were well short of what would ultimately be required to "beat" global warming. Nevertheless, as minimal as the negotiated reductions were, those negotiations were reportedly intense and contentious.

So, the question is this: if a wholly green future, free of fossil fuels, really is a lower cost energy option as many climate change warriors claim, why aren't the world's representatives at these climate conferences falling all over each other to commit to massive emissions reductions? Why wouldn't every nation on earth be clamoring to be the first nation to become a zero emitter? If wind, solar and other green alternatives are so low cost, not only would those nations become carbon free, but their economies would also presumably surge forward riding the wave of lower cost energy! Their businesses would enjoy an energy cost advantage that would allow their economies to soar! In fact, why would a negotiation even be necessary? Any nation that wanted to prosper could unilaterally choose to race down this least-cost, green-energy pathway, right?

Well, that's not happening. Do you think maybe those climate-action diplomats know how expensive and how disruptive it would be for their nations to turn their backs on fossil fuels? Do you think maybe the intensity of the negotiations reflects the fact that they know that the green future that they and we are all trying to achieve and the world so desperately needs, wouldn't be just a little bit more expensive, but maybe a lot more expensive? My point is that green advocates (like me) need to be honest and straightforward: saving our planet, if we limit ourselves to wind and solar power and want to enjoy the same level of electric system reliability we now have, is going to significantly increase the cost of energy, and everything else, period.

Finland's Prime Minister Sanna Marin was featured in the TIME magazine Feb. 3, 2020 edition and was quoted as saying: "I don't think that fighting climate change means higher costs and a worse future. I think it's the opposite."

On the cost question, I'm sorry, but for all the reasons I've just enumerated, I think she's wrong. And anyway, if the Prime Minister really believed what she was saying, what's preventing Finland from ending its consumption of fossil fuels immediately? If it wouldn't mean higher costs and a "worse" future, then why wouldn't Finland simply embrace the Nike motto and "Just do it?" She could use all of the energy cost savings Finland would realize by abandoning fossil fuels to reimburse her fellow Finns for replacing their internal-combustion-engine-powered cars and trucks and all their other fossil-fuel-burning durable equipment. Why not just prove to everybody that talk isn't the only thing that's cheap?

As to the issue of whether or not a future that eliminates fossil fuel is "worse," I'm not sure I know what that means. Worse

> *A future without fossil fuels would be different and needs to be different if we have any hope of limiting the environmental damage we will cause if we don't stop emitting CO_2.*

(or better) in what ways? Worse (or better) for whom? I don't think a values-laden judgment as to whether it's worse or better is particularly relevant. I think Prime Minister Marin would certainly agree that a future without fossil fuels would be different (and needs to be different) if we have any hope of limiting the environmental damage we'll cause if we don't stop dumping CO_2 into the atmosphere.

Incidentally, the largest single energy source in Finland is wood. Their second largest energy source is oil.

In the course of researching this book, I've come across a number of optimistic forecasts that echo the outlook of Prime Minister Marin. I'll highlight just one of them as representative of the genre: Mark Hertsgaard's 2011 book, Hot: Living Through the Next Fifty Years on Earth. The author writes:

> "Happily, there are genuine reasons for hope. Not only do we know what it will take to stop global warming, but most of the necessary technologies and practices are already in hand. Best of all, putting these tools to work could actually strengthen our economy, improve our quality of life, and make money, lots of it."

I agree with Mr. Hertsgaard that we have many of the tools necessary to tackle the global warming challenge, but we've had most of those tools for two decades now, over which time GHG emissions have not only not been going down, they've been going up at an increasing rate. And yes, decarbonizing the world energy economy will generate massive numbers of jobs and economic activity. But all of those jobs and all of that economic activity would have the sole function of replacing our existing energy infrastructure, not increasing "productivity" for real growth. Instead, we will be substituting skilled labor and capital investments to replace incredibly energy-dense and comparatively inexpensive fossil fuels. Making that substitution will raise energy costs and that's why the gentle lady from Finland and governments all around the world (including ours) have talked a lot about curbing global warming but haven't taken the unilateral steps that they know would damage their economies.

Have we made progress from where we were 30 years ago? **Yes,** absolutely. Have we made enough progress to warrant being truly hopeful? **No**, not by a long shot.

A Lingering Legacy

Before leaving the issue of how expensive it's going to be to transition away from fossil fuels, it's worth noting that it's not just old coal plants that will have to be shut down in order to reduce worldwide CO_2 emissions. New coal plants are still being built around the globe **today**. In the recent past, China was reportedly building one new 500 MWe coal-fired power plant **every week.** A 500 Mwe power plant is capable of serving the power needs of between 500,000 and four million people depending on the intensity of their energy use. As astounding as these numbers are, this rate of adding coal-burning electric generating capacity is half the rate they were building in the decade from 2005 to 2015. All of their new coal plants are not only going to increase CO_2 emissions as they come online, they'll continue emitting CO_2 for the design lives of the plants, typically at least forty years. The operators of those plants, especially the newer ones, are not going to willingly abandon those investments and leave their customers in the dark.

But it's not just the generation of electricity where there are economic ramifications of transitioning away from fossil fuels. The transition to a carbon-free economy will gradually devalue energy-related mineral assets. A country with vast coal reserves or a mining company that owns fossil-fuel mineral rights is going to fight hard to allow fossil fuels to continue to be burned. The owners of the assets will try to preserve their value. Consequently, the energy marketplace during an era when aggressive actions are being taken to reduce CO_2 emissions will be dynamic, but it will also include many players who will attempt to prevent change and sustain the status quo.

Coal will probably devalue the fastest in countries that have abundant natural gas. That's what's happening now in the U.S. Oil will probably devalue only gradually as the world demand for petroleum begins a very long, very slow slide, not in this decade maybe, but possibly by 2050. As the value of oil declines and the oil industry becomes less profitable, we run the risk of increased environmental damage if the oil infrastructure (wells, pipelines, storage facilities, etc.) aren't maintained as well as

> *The transition away from fossil fuels will not only be expensive, it will also be slow and contentious.*

they need to be. In general, national and local interests will come into conflict with global emissions reduction efforts. These interests will give rise to powerful forces to resist the changes necessary to combat global warming. Making the transition away from fossil fuels will not only be expensive, it will also be slow and contentious.

A Carbon Tax

I touched on the issue of carbon credits in the chapter on Transportation. As you already know, I'm not a fan.

Proponents of such schemes argue that they're better than nothing; **that may be true, but I'm not sure.** The problem is these approaches still allow companies that emit large quantities of CO_2 to continue operating and producing their products. I guess you could say the "clean" companies that are selling them carbon credits are "enablers."

It makes sense to provide incentives for companies to clean up their manufacturing and their products, but in terms of protecting the planet from global warming, it makes no sense to then give back some or all of those environmental benefits to high emitting companies and companies that sell high-emitting products. We don't offer incentives for people to install solar panels on their rooftops so that the solar panel owner can sell credits to their neighbors allowing them to drive a diesel pickup truck to the store to buy milk.

I'm particularly skeptical of forestry related carbon credits. One of ways companies and governments can earn emission reduction credits is to plant trees or protect existing stands of trees from being cut down. I love trees and support tree-planting almost anywhere, but I seriously question the value of using tree-planting and protecting portions of existing forests as a strategy to allow other profit-making enterprises to continue emitting CO_2.

To our great horror, we've all observed, seemingly unstoppable fires destroy great swaths of forests in a matter of weeks or months. Forestry related carbon credits may give us false comfort that we're making great strides. Personally, I'd much rather see real emissions from all human activities be driven down to zero without using carbon credits and instead, leave tree-planting and forest protections (which we should be doing anyway) as mechanisms to draw down excess CO_2 concentrations in the atmosphere **on the way to** and **after** we get to "real-zero."

In general, I'm not in favor of introducing artificial incentives or disincentives into the marketplace, i.e., subsidies or taxes, because they distort the free market and if they're not done carefully, they can send erroneous pricing signals to consumers.

However, in the course of writing this book I've come to the conclusion that a carbon tax (or

> *A carbon tax (or something like it) is essential to combating climate change and is the most straight-forward way of including the cost of CO_2 emissions in the current price of energy and everything else we consume.*

something like it) is the single most important thing that could be done to dramatically alter the global warming trajectory we're on. It's also the most straight-forward way of including the cost of CO_2 emissions in the current price of energy and everything else we consume.

A carbon tax on fossil fuels is like a sales tax on consumer purchases. It's an extra charge added to the transaction and paid by the consumer. Levying a carbon tax is a way of incorporating the cost of CO_2 emissions in the price of goods and services, i.e., a way of accounting for the "externality" of carbon emissions that had previously been ignored.

Aside from raising revenue, the whole purpose of a subsidy or a tax is to influence behavior. Subsidies generally lower the cost of a purchase for consumers and are intended to positively influence their decision to buy that product, e.g., rebates for solar panels, dual pane windows, and high efficiency appliances. Taxes, on the other hand, raise the cost of a purchase and are intended to discourage the purchase of those items, e.g., liquor and tobacco products.

In principle, a carbon tax is fairly easy to explain, but probably more complicated than I can imagine to implement fairly and once implemented, to enforce. The fact that there may be many as-yet unanswered questions does not mean we shouldn't pursue it. We need to start immediately and refine the process as we go along. We can't let the "perfect be the enemy of the good."

Some of the questions that immediately come to mind are:

1. Where in the carbon consumption cycle should the tax be applied: at the point of resource extraction from the ground, at the refinery, at the point of sale, at the point of consumption, at all of these?
2. How much will the tax be (i.e., \$/ton CO_2)?
3. Will the tax be progressive, i.e., higher for high fossil-fuel users and if so, how will it be determined who the excessive-consumers are?
4. Who will collect the tax revenues?
5. Will the tax be equal in all jurisdictions? Will taxes be higher in countries or regions with higher CO_2 emissions per capita?
6. How will the tax revenues be used? Will some of the revenues be funneled back to low-income individuals to partially offset the regressive nature of a flat carbon tax? Will the revenues be used exclusively for fossil fuel and GHG reduction programs?
7. How will the tax be implemented across jurisdictional lines, i.e., regions, states, countries?
8. If all countries don't implement the tax, what mechanisms will be used to "punish" non-participating countries?
9. Will all sources of CO_2 emissions be taxed equally, e.g., cement production vs. fossil fuels extracted from the earth?
10. Will processes that have a viable carbon-free alternative be taxed more than processes that don't?
11. How will compliance be monitored and by whom?
12. Will a carbon tax be applied on top of existing taxes such as gasoline taxes?
13. How will a carbon tax be levied on leakage from the natural gas and oil distribution systems and on refinery flaring?
14. Will the "carbon tax" be higher on GHGs with higher GWPs (Global Warming Potentials) such as methane?
15. Farm animals are a significant source of methane emissions. Will a "carbon tax equivalent" be applied to farm animal "emissions?"

While there are sticky questions about how to implement a carbon tax, one of the most important things this approach has going for it is that **it allows the market to find the most efficient and the most effective way to cut emissions in a way that no other policy lever can.**

Exhibit 13.1 provides a list of some of the most commonly discussed strategies and programs to address our CGI. All of these approaches will be used to some degree or another to reduce carbon emissions in the coming decades.

How can we know which of these approaches or what combination of approaches will yield the "biggest bang for the buck" in terms of

> *How can we know which of the GHG reduction strategies or combination of strategies will yield the "biggest bang for the buck?" We can't, but <u>the marketplace can</u>.*

driving down emissions? Also, how can we know if that "bang" will be the same everywhere in the world? The answer is, **"we can't know in advance which strategies will be most cost effective, but the marketplace can."** And "No! The answers are not the same everywhere in the world."

As we ratchet up the carbon tax, the activities and products that emit the most GHGs

> *A carbon tax allows the market to find the most efficient and the most effective way to cut emissions in a way that no other policy lever can.*

will become increasingly expensive first and people will shy away from the more-polluting, more-expensive products and activities. In theory, they'll find the less polluting, less expensive alternatives more appealing. Used Hummers will become a less attractive transportation option while all-electric pick-up trucks may find some new, willing buyers, all without the government making judgments about which technology or product is "best." Additionally, once in place, the carbon tax can be adjusted up or down to achieve whatever level of emissions reduction is deemed necessary.

There are many across the political spectrum who rail against "taxes." I would urge them to recognize that a carbon tax is not a

Exhibit 13.1
Climate Action Toolbox

- Solar Panels
- Wind Turbines
- Biomass
- Hydroelectric Generation
- Pumped Hydro Storage
- Chemical Battery Storage
- Carbon Capture
- Energy Efficiency
- Conservation
- Tree Planting
- Industrial Process Improvements
- Cloud Brightening
- Soil Management and Enhancements
- Reduced Consumerism
- Wetlands Protections
- Forest Protections

- Mass Transit
- Urban Planning
- Rail Electrification
- Appliance Electrification
- EVs
- Plant-Based Meats
- Geothermal Generation
- Nuclear Power
- High Voltage Transmission
- Packaging Minimization
- Recycling
- Material Substitutions
- Lifestyle Changes
- Carbon Tax
- Innovation #1
- Innovation #2
- Innovation #n

revenue-generator designed to expand government. On the contrary, it's a tax designed to reorder the marketplace and change behaviors. Each of the strategies or technologies will compete with one another. The consumer will render the ultimate decision for goods and services while policy makers will compare programs with different cost/benefit returns – in each case with different outcomes because of the presence of the carbon tax. It's actually an approach that has the potential to shrink government. The marketplace would pick "winners" and "losers," not the government. The power of the market would be unleashed to find the most efficient and effective path forward without any additional bureaucratic or regulatory intervention.

Recognizing that global warming is an existential threat to the nation (and the world) and to the stability and success of nation states, a carbon tax could also be thought of as an element of "defense spending." If the tax contributes to lessening the impact of global warming, some of the more-destabilizing geopolitical factors may also be lessened.

In order to avoid the dreaded "tax" word, if need be, the carbon tax could be renamed as the "Planet Survival Investment" charge, a PSI. Anyone "liberating" carbon from being tied up in a fossil fuel would be required to make a deposit in a PSI fund.

If it's to be successful in bringing CO_2 emissions down, the carbon tax, or PSI charge, will have to be universal. It will have to apply to all states, all regions, and all countries equally. If it doesn't, the market for goods and services would not be a level playing field. Countries refusing to implement a carbon tax would have a lower cost of energy and a competitive advantage compared to the participating countries. The lack of equity would in turn encourage cheating and manipulation even in participating countries. Order and equal enforcement would break down.

Short of obtaining 100% participation by all countries, a carbon tax could be made effectively universal if a sufficiently large fraction

> *People don't like to change and they especially don't like being forced to change.*

of the countries making up the world economy agreed to adopt the tax. Those countries could make the tax universal by agreeing to impose trade and travel restrictions on non-participants.

Whatever it's called, a carbon tax or a PSI charge or whatever, when it's implemented, it's likely to be low at first so that people can get used to the concept. However, in order for the tax to be effective, it must ultimately be high enough to influence behavior. For a carbon tax to be truly effective i.e., dramatically reduce the amount of CO_2 being pumped into the atmosphere, eventually it needs to be fairly high. It needs to be high enough to disrupt existing consumption-oriented lifestyles. If it isn't high enough to cause a change in behavior, it won't be effective. However, not surprisingly, people don't want their lifestyles disrupted, so

> *People don't want their lifestyles disrupted. As a result, it's highly likely that there will be a lot of opposition to a meaningful and effective carbon tax, even for a worthy cause like saving the planet.*

it's highly likely that there'll be a lot of opposition to a meaningful and effective carbon tax, even for a worthy cause like saving the planet.

I address the obstacle reaching a broad, international consensus in the next two chapters).

National and local economies will also be disrupted as the cost of energy increases. Economies that rely on the movement of goods and people, will have to be replaced with something else. The allocation of household budgets for energy will increase and the funds available for discretionary spending will shrink. Travel will become more expensive. Tourism and all the jobs associated with it are likely to take a significant hit. The mass movement of farm produce around the world (e.g., strawberries for Christmas in New England) is also likely to be impacted.

Competing Demands for Limited Resources

In the years ahead, climate experts warn us that extreme weather events are likely to become more frequent and more severe. Nearly all of us have noted disruptions to normal weather patterns including too much precipitation in some places, too little precipitation in others, higher peak summertime temperatures and higher nighttime low temperatures almost everywhere. All of these changes have economic consequences. As the effects of global warming cause changes in regional climate characteristics and the impacts of weather events become greater and greater, the competition for limited resources will intensify. Governments (and taxpayers) will have to simultaneously apportion limited resources in three distinct and competing areas:

1. Rebuilding and supporting communities after weather-related disasters (e.g., floods, tornadoes, hurricanes, droughts, forest fires, famines, community/population relocations),
2. Investing in infrastructure to protect existing investments (e.g., seawalls and levees) to protect farms, ports, cities, and industrial enterprises at least temporarily, and
3. Investing to replace the existing fossil-fuel-based infrastructure (i.e., power generation and energy storage).

When faced with the need to make expenditures to help people recover from disasters, longer term initiatives (like replacing fossil-fuel-burning power plants and electrifying railroads) will be postponed. It'll always be more convenient (and politically acceptable) to postpone making tough decisions and long-term investments, especially if people need food, water, and shelter after a disaster. This is what I mean by competing demands for limited resources. If protecting the future of the planet comes into direct conflict with the immediate needs of the world's current population the "here-and-now" will always take precedence. The future will nearly always draw the short straw.

> *The "here and now" will nearly always take precedence; the "future" will always get short-changed.*

Recovering from, preventing, and eliminating the causes of disasters are all compelling areas in which to make expenditures. All three will rise over time. These increased expenditures, however, will increase the strain on financial systems. As times get tougher, most people will have fewer resources, less time to devote to discretionary activities and less tolerance for anything that hints at making their already-shrinking standard of living shrink even faster (no matter how high or low that standard of living may be).

Mini-Recap

Money makes the world go 'round. Those aren't just catchy lyrics to a Broadway show tune; they're a fact of life.

CO_2 emissions are an economic externality that's been ignored for decades and is only now being re-introduced into energy analyses. Bringing this externality into the economic equation is necessary to combat global warming but will make energy more expensive in the future.

Serious efforts to reduce GHG emissions will have profound economic ramifications beyond the electric generation market. Some mineral assets will be upvalued while others will be devalued. Similarly, some infrastructure assets will be upvalued while others will be devalued. The same impacts apply to capital equipment investments, durable goods of all kinds and even land. The point is this: all economic activities, worldwide, are intimately linked to our intensive dependence on fossil fuels. Transitioning our energy economy to carbon-free resources means restructuring the entire world economy.

> *Transitioning our energy economy to carbon-fee technologies means, restructuring the entire world economy.*

Increasing energy prices will squeeze all consumers. All of us will have to pay more for the energy we consume directly and for the energy content in everything we buy. Governments around the world will have to find a balance between responding to weather-related disasters, fortifying existing infrastructure to prevent additional near-term disasters and making the basic infrastructure investments to decarbonize their energy sectors. A carbon tax could be an efficient and effective way of moving the world to a carbon-free future and providing some of the funding to support such a move, but a visible and effective carbon tax will be difficult to both implement and enforce. Opponents will vigorously attack a carbon tax (or even the suggestion of one) as economic suicide, conveniently ignoring the fact that if a carbon tax is applied universally, no country or state will have a competitive advantage over its neighbor.

Action Items

Economics is the key to minimizing damage to the planet. Here are a few of the economic levers that could help address climate-related stresses.

1. Establish a worldwide PSI charge, i.e., a Planet Survival Investment charge (aka, a carbon tax) that increments upward over time. Impose an equivalent PSI on agricultural and industrial sources of GHGs.

2. End "cap and trade" and "regulatory carbon credit" programs. Until they're eliminated, make the financial benefits of green products flow to the product owners and not the product manufacturers (e.g., EV owners, not EV manufacturers).

3. Terminate all U.S. coal exports within five years.

4. Privatize flood and disaster insurance. After disasters, limit government aid to search and rescue. Get the government out of the business of helping people rebuild in areas prone to weather-related damage (fire, floods, and excessive heat).

14 ATTENTION, ACCEPTANCE, ALIGNMENT AND ACTION

"What[ever] you do or dream you can, begin it;
Boldness has genius, power and magic in it."

Goethe, Faust, 1829

The process of moving people from complacency to action on the issue of global warming involves a sequence of steps:

1.   First, getting people to focus on the issue (i.e., attention),
2.   Second, getting people to understand the issues and embrace the science (i.e., acceptance),
3.   Third, getting a critical mass of people onboard to build a consensus about what needs to be done (i.e., alignment), and
4.   Following through on the plan (i.e., action).

Attention

It's really hard to get people to focus on anything for any extended period of time these days. We live in a world full of stimuli coming at us quite literally 24/7. There are so many issues competing for our eyeballs – and they're all (literally) a finger-touch or voice command away. Exhibit 14.1 lists just a tiny fraction of things that might be "competing" for your attention. [Note: this exhibit is clearly "dated." By carefully examining its contents you could probably pinpoint to within a few months the date I originally compiled it. I contemplated updating it to include more current concerns but concluded that this "outdated" list actually serves to reinforce the point I'm making – that there's an **ever-changing and ever-expanding** set of issues screaming for our attention.]

Exhibit 14.1 What's On Your Mind?

Endangered Species	Mass Shootings	Wealth Inequality
White Supremacy	Gun Rights	Democracy
Melting Glaciers	Terrorism	Noise Pollution
Natural Disasters (floods, tornadoes, hurricanes, fires, volcanoes, earthquakes, etc.)	Man-made Disasters (airplane crashes, building collapses, rail crashes, ship sinkings)	Compensation for College Athletes
Me Too	Organic Foods	Pandemics
Animal Rights	Opioids	Unemployment
Voting Rights	Vaping	Self-Driving Cars
Refugees	Famine	Education Reform
Organ Transplants	Domestic Violence	Travel
Privacy Rights	Consumer Protections	Authoritarian States
Drug Cartels	Interest Rates	Trade Wars
Tariffs	Vaccinations	Robo calls
College Applications	Rogue nations	Breaking News
Culture Wars	Insider Trading	Water Quality
Hackers	Test Scores	Fake News
Regulations	Coral Reefs	Social Security
Aging Parents	Family Leave	Black Lives Matter
Save the Whales	Earthquake Insurance	Olympics
Fraud	Plastics	Doping
Minimum Wage	Minority Rights	Water Rights
Crime	Password Protection	Covid-x
Police Brutality	Housing Costs	Tax Reform
Asteroids	Bitcoin	Sleeplessness
GMOs	Fireworks	Mold
Evolution	Internet Access	Grandchildren
Labor Standards	Nationalism	Immigration
Wildfires/ Forest Fires	Sentencing Reform	Food Safety
Space Junk	Veterans' Care	**And on, and on**
Cancer Cures	Homelessness	**and on, and . . . . . .**

We're all interconnected, often more than we want to be. When deprived of our phones or computers some of us discover that our need for an electronic connection is like an addiction. Some people feel physically ill when their electronic connectivity is taken away. While "anytime, anywhere" personal connectivity has yielded many benefits, I worry that it also has downsides. I worry that people are (becoming) so connected with the outside world that they are becoming disconnected from their inner selves.

As an aside, I also worry that constant outside stimulation is making us less creative. Creativity requires that our brains have some down time to regroup, reshuffle and reorganize experiences. I'm afraid that constant stimulation interferes with this critical aspect of creativity . . . but I digress. [See what I did there? I'm writing a book about global warming and I got distracted. I wandered off talking about connectivity and creativity in a chapter all about **focusing** . . .]

All these outside distractions/issues/concerns (and countless others) demand our attention along with everything else we have to do and take care of. Is it any wonder that most of us can't focus our attention on any one issue for any extended period of time?

"Beating" global warming, however, is going to take a single-minded, laser-focus that lasts for decades. With so many issues vying for a place on the never-ending hit parade marching through our brains, it would take a miracle for this one issue to be foremost in our collective psyche for even a day, much less over a period of decades and across generations.

Extreme weather events will help refocus our attention on global warming from time to time, but in some ways, a constant drip of climate change warnings punctuated by significant weather-driven disasters may actually work against decisive, sustained action. Just like the proverbial frog that tolerates a slowly heating pot of water until it boils to death (a metaphor, by the way, ungrounded in reality), a steady stream of climate change warnings may become so routine that it'll be ignored. Perhaps that's already happened.

Acceptance

Assuming you can get people to focus on the issue of global warming for any extended period of time, people still have to understand the problem, their role in solving it, the impact of mitigation measures on them personally and the consequences if the planet's heat balance isn't restored. I'm often surprised to hear intelligent people latch onto the arguments of one or two outliers in the global warming debate. They reject, without explanation, the data compiled by the vast majority of the scientific community and discount the consequences they can see and feel with their own senses. They are entitled to their opinions, but they should at least be required to provide an alternative explanation for the totality of the evidence they're rejecting.

Alignment

Getting individuals to focus on the issue of global warming, process the data, understand the science, and accept that it's real and urgent are all prerequisites to building a group consensus. As more and more individuals accept the reality and urgency of global warming, the next step is to build a consensus around what needs to be done -- achieving, what I'm calling, alignment.

How a group of people feels about an issue can be represented as a collection of vectors. A vector has two characteristics: a direction and an amplitude. The direction of a vector can be thought of as an opinion about an issue and its amplitude (length or size) can be thought of as the number of people who hold that opinion or the intensity of feeling supporting that opinion. A collection of vectors, then, can be used to visualize how a population feels about an issue.

Consider the issue of taxes. A person who strongly favors taxing corporations and wealthy individuals to fund government programs could be characterized by a long arrow (a vector) pointing to the right.

Exhibit 14.2 Vector Representing "In Favor"

Folks who strongly oppose higher taxes on corporations and individuals could be characterized by a long arrow to the left.

Exhibit 14.3 Vector Representing "Opposed"

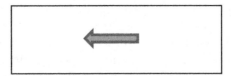

And maybe there are some who favor taxing corporations heavily but not individuals or vice versa. These folks might be represented by vectors pointing to "2 o'clock" or "8 o'clock." How a population feels about taxes can be represented with a collection of vectors like that shown in Exhibit 14.4. This exhibit provides a visual representation of how a population feels about an issue on which there is no clear consensus. The opinion vectors are not aligned.

Exhibit 14.4 Vector Representation of an Issue
With No Clear Consensus

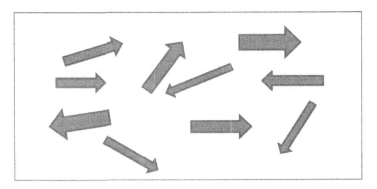

In contrast, Exhibit 14.5 is a visual representation of an issue on which there is broad agreement; the opinion vectors are aligned.

Exhibit 14.5 Vector Representation of an Issue
With Broad Agreement

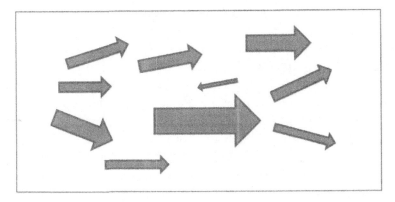

Such a representation might characterize how the country felt about entering World War II after Pearl Harbor or retaliating against Al-Qaeda after the terrorist attack on 9/11. (Note that even on issues where there's a broad consensus, there's still likely to be some dissent).

The extent to which the opinion vectors are aligned shows the strength of the consensus. The greater the alignment, the more likely the group can coalesce around a set of responses. Complex issues (like global warming), however, seldom, if ever, have an opinion matrix that's aligned as totally as that depicted in Exhibit 14.5.

Lights, Camera, ACTION!

After focusing the nation's attention on the issue of global warming, acquiring a deep understanding of the issue and reaching a broad consensus, there's still no guarantee that there'll be any action. As described in the previous chapter, vested interests will campaign hard to maintain the status quo that favors their financial or power-based interests. Large corporations and well-funded, politically active minorities have thwarted the will of the majority on any number of issues. At the turn of the century food manufacturers resisted the pure food movement and government regulations proposed by Dr. Harvey Wiley, first commissioner of the U.S. Food and Drug Administration. In the early 1960s, the chemical industry attacked Rachel Carson when she exposed the environmental impact of DDT. In the 1960s, '70s and '80s, the tobacco companies deliberately played down the damaging health effects of smoking. Today, oil and other fossil fuel interests have successfully opposed improving car mileage standards and virtually all carbon-reduction policies.

On the issue of global warming, Exhibit 14.5 may represent how the country is aligned in terms of their concern about global warming. Exhibit 14.4, however, may be a better representation of what the country thinks we should do about it. It's doubtful that we'll ever achieve the degree of alignment represented in Exhibit 14.5 on a plan to aggressively reduce CO_2 emissions. As the consequences of global warming become more severe, it's reasonable to expect that the alignment of public opinion to address the problem will increase, but only marginally and probably not consistently.

Moving forward with an all-encompassing program to dramatically reduce GHG emissions will take not only nearly complete alignment across the population of one country, but also across the populations of virtually all countries over a period of literally decades. (The international dimensions of this challenge are discussed in the next chapter). Those actions will be expensive and sometimes unpopular and will give rise to intense arguments. In democratic countries, politicians will be under intense pressure to relieve the immediate burden on their constituents and their economies. Because few politicians will still be in office decades into the future, many are likely to take the expedient path and weaken or delay CO_2 countermeasures rather than risk being voted out of office. Modest programs, i.e., those with the smallest price tags and least consequential impacts, will continue and will be heralded as great successes, but modest programs won't "beat" global warming.

Any meaningful effort to halt global warming is going to require momentous changes in behavior. We American can make significant cuts in our personal carbon footprints without sacrificing much comfort, but in spite of that, we, like people everywhere, won't react well if someone or some policy tries to take away something (whatever that something is) that we currently have. Because climate change countermeasures will almost always take something (either money or choices) away from affluent consumers, Americans may react badly if they feel their personal choices are being limited. That's going to make it that much harder to develop and sustain an action-oriented consensus over an extended period of time.

The other thing about us Americans is that we're not overly patient. We're not good at deferred gratification. We want action right now on whatever issue

> *We Americans are not good at "deferred gratification."*

confronts us. We want and expect immediate rewards. But combating global warming is all about deferred gratification. Building consensus is hard enough when the rewards come quickly, but doubly hard when the proposed actions are "painful" in the present and the rewards are both difficult to quantify and are predominantly far in the future.

Consider our experience attempting to increase gasoline taxes – a public policy decision that could reduce gasoline consumption, reduce CO_2 emissions and improve our balance of payments. Regrettably, there's been painfully little public support for higher gasoline taxes in the U.S., even when the added revenues are earmarked for road repairs.

Given the experience with the gasoline tax, how can anyone be confident that the public will endorse financially painful policies today to achieve benefits in the distant future?

Public support for measures to combat climate change will ebb and flow. Major weather events will heighten concerns and promote actions, but both will subside as the memories of the events fade . . . until the

> *There's no way we're going to get and keep the public's <u>attention</u>, <u>agreement</u>, and <u>alignment</u> to take the <u>actions</u> necessary to "beat" global warming.*

next one. But this "on-again, off-again" approach won't get the job done.

Many people say they support action to combat global warming, but when it comes down to real actions in real time, support seems to melt away like a polar ice cap. **I won't be convinced that there's a meaningful groundswell of support for policies and actions to combat global warming until Americans voluntarily open their wallets to save a future they'll never personally experience.**

Unplugging from fossil fuels will be somewhat like having a root canal: expensive and painful, but necessary in the long run. The similarity to combating global warming is that fixing the problem will

> *Actions to effectively combat climate change would be like having a root canal that lasts for decades so that someone who hasn't yet been born may feel modestly better decades or even centuries in the future.*

also be painful and expensive, but the similarity ends there. The root canal patient has to endure a sometimes-painful, hour-long experience to derive the benefit of the procedure, whereas the pain of combatting global warming will last for decades and the primary benefits will flow to generations who haven't yet been born.

If we push as hard as we need to in order to get done what needs to get done, the "alignment" necessary to support those policies and programs will be shattered. If the pain becomes too great, "NIMTOE-ism" (Not In My Time On Earth) will take hold.

> ### *"Catch 22"*
> *If we push as hard as we need to*
> *to get done what needs to get done,*
> *the alignment necessary*
> *to support those policies*
> *would melt away.*

Would We Do It If We Could?

I've argued that the rapid transition away from fossil fuels, if we could do it, would be expensive. One measure of that expense is in dollars and cents, but the transition would also completely upend lives and livelihoods of the citizens of every country in less than one generation. The economic order that's in place now has developed and evolved over decades. Take tourism and leisure travel for example. If travel becomes so expensive that it's inaccessible to the average person, the travel industry will shrink dramatically. Airlines, cruise lines, travel companies, hospitality enterprises, hotels, food service – they'd all shrink. That's why resistance to global warming countermeasures will grow. At the heart of that opposition are the people whose lives will be impacted by efforts to counteract global warming. The ripple effects through every aspect of the economy will be magnified and become tsunamis. We simply couldn't handle the resulting economic mayhem if we somehow garnered the courage necessary to act decisively to combat global warming.

A Tightrope

Policy makers, regulators and negotiators have an impossible dilemma. They will try to develop actions aggressive enough to take a bite out of CO_2 emissions, but not so aggressive as to lose the support of the people who will have to live with those new policies. Unfortunately, what needs to get done to "beat" global warming lies well beyond any viable compromise.

15 CO_2 WITHOUT BORDERS: INTERNATIONAL COOPERATION

**"If you want to go fast, go alone.
If you want to go far, go together."**

African Proverb

I've been making the case that abandoning fossil fuels will require an enormous logistical and economic effort and that powerful forces "prospering" under the current status quo will seek to prevent that from happening. I've also argued that abandoning fossil fuels will dramatically increase energy costs which will weaken the world's resolve to confront the problem in any meaningful way. None of these issues, however, has yet touched on probably the most critical requirement to address our Catastrophic Global Imbalance (CGI): **international cooperation.**

If each country controlled its own atmosphere and heat balance, global warming would be a fundamentally different issue. Nations with a long-term outlook and educated populations that understood the threat could take the necessary steps to ensure their own survival. They could make the investments and adopt the policies to avoid catastrophic local weather events. Nations with a short-term perspective or ones that couldn't convince their population to make the

> *If each country controlled its own atmosphere and heat balance, climate change would be a fundamentally different issue than it is. Unfortunately, it doesn't work that way.*

changes necessary to avoid catastrophic weather events, would alone suffer the consequences of their short-sighted policies.

Unfortunately, it doesn't work that way —which, of course, is why we call it **global** warming.

Our atmosphere and our oceans don't respect the national boundaries we've drawn on maps.

1. Ocean currents carried debris from the Fukushima area to Oregon in a matter of months. These same currents concentrate plastic debris from all over the Pacific in several large Pacific Garbage Patches.
2. Air currents carried radioactive isotopes from Chernobyl far beyond the destroyed plant. Radioactive by-products from atmospheric atomic bomb testing in the 1950s and '60s were detected all around the globe.
3. The volcanic ash from the 2020 eruption of the Eyjafjallajökull volcano in Iceland forced cancellation of the flights of some ten million airline passengers across Europe.
4. Tiny combustion particles from the bush fires in Australia settled on the ice sheets in Greenland.
5. In the aftermath of the 9/11 terrorist attacks, virtually all commercial aviation was grounded. Air quality improvements were detected all around the globe.
6. In late 2019 and continuing into 2022, COVID-19 spread around the globe in a matter of months ultimately killing millions and disrupting commerce.

Like it or not, we are all, quite literally, "in this together." We are all interconnected.

> **Global warming is the ultimate form of globalization.**

Global warming is just the ultimate form of globalization.

This simple fact presents the most challenging aspect of efforts to combat global warming and the climate change that will flow from it. It's also the reason international meetings have been and will continue to be convened to address the issue. And it's why those conferences have been so contentious and disappointing in terms of real progress.

The delegates to those conferences grapple with the same kinds of questions that were laid out in the discussion of a carbon tax:

1. How much will each participant nation have to cut back emissions and how quickly?
2. Should the emissions reductions be voluntary or mandatory?
3. If they're voluntary, will they work?

4. If they're mandatory, how will monitoring be done? Who will do the monitoring?

5. How will violators and/or non-participants be "punished" and by whom?

6. What are the obligations of richer nations (which typically have higher per capita emissions) compared to poorer nations?

7. Who will pay to relocate displaced peoples from low-lying or island nations? Who will accept these climate refugees?

8. Who has the authority to negotiate?

CFCs Part 2?

The world response to the appearance of a hole in the ozone layer and the subsequent banning of CFCs is often used as an example that international cooperation to curb CO_2 emissions is possible and could be successful. The argument is made that since the world was able to come together to close the ozone hole, we should be able to do it again to "beat" global warming.

I respectfully disagree.

The worldwide action to reduce or eliminate chlorofluorocarbons releases was an important environmental action to be sure, but the suggestion that the CFC ban can be replicated and expanded to "beat" climate change is, I believe, a willfully optimistic mischaracterization of the issue. The agreement to ban CFCs was comparatively easy because modestly priced chemical substitutes for CFCs existed. Most consumers were virtually unaffected by the transition away from CFCs. With apologies to the negotiators who worked hard to negotiate the actual agreement, the CFC agreement appears to have been almost a "no-brainer."

Certain factories had to change processes, but nobody had to give up their air conditioning or refrigeration. The modest extra cost was paid by those using the replacement refrigerant making it relatively painless to abandon a product that was doing immediate and measurable harm to the physical environment. It was a comparatively inexpensive, quick, technical fix. In contrast, disconnecting from fossil fuels will be hard, expensive and take a long time.

Any effective international agreement to limit CO_2 emissions will have to include every nation (or at the least, all of the big emitters). Attempting to significantly reduce CO_2 emissions without having every major emitter on board would be a little like trying to carry water in a bucket that has a big hole in it. Any agreement that's not fully inclusive couldn't possibly accomplish what's required, wouldn't be able to do it quickly enough and would pretty quickly fall apart.

Non-participants (i.e., countries still burning fossil fuels) would have a cost advantage that would allow their products to be more competitive (or more profitable) than those from participating countries. Disadvantaged countries would have a tough time convincing their citizens that they needed to sacrifice for the good of the planet while others were profiting. Whether the approach to reduce worldwide CO$_2$ emissions is an enforceable multi-party agreement or a carbon tax, broad international cooperation will be required.

And it's not just the rich nations with high per capita CO$_2$ emissions that have to cut back; all major CO$_2$ emitters will have to reduce their emissions.

Who Will Cut and by How Much?

What matters in the effort to slow the rate of global warming is **total emissions**. It's the total quantity of CO$_2$ being pumped into the atmosphere that has to be drastically reduced or eliminated. Countries with high per capita emissions but comparatively lower populations (e.g., the U.S.) have to reduce emissions, but so do very populous countries like China and India with lower per capita emissions but such large populations that they are still emitting huge quantities of CO$_2$.

The issue of per capita CO$_2$ emissions provides some insight on the issue of how much each country might need to reduce emissions and how hard it will be to achieve those reductions. Countries with the high per capita CO$_2$ emissions (such as the U.S.) have a moral responsibility to the rest of the world to achieve the largest percentage drop in the shortest possible time. Fortunately, the U.S. also has the financial resources that could make this happen (if a consensus develops) and can make many of the initial cuts without overly adverse effects on the quality of life that most Americans enjoy. In the U.S., we have a lot of fat to trim (literally and figuratively) before we cut into muscle. After those initial easy cuts, the subsequent cuts will be larger and more disruptive.

Populous countries such as India and China are particularly problematic. How do you deny the citizens of a country with a low per capita annual income AND low per capita CO$_2$ emissions access to the benefits (luxuries and necessities) of low-cost energy? Solar panels could bring intermittent but at least comparatively low-cost energy to literally millions of people who currently have no access to any power grid. Recipients of this newly available energy source will consider it a godsend! But solar panels alone can never bring the level of prosperity and wealth generated in the U.S. and Europe by the energy bonanza that was at the heart of the industrial age. India and China also have large reserves of coal which presents the additional ethical dilemma: how do you deny the citizens of these countries access to domestic energy supplies that they so desperately need? How do you tell them not to exploit the same resources that were used by the West to build its industrial base (and wealth)?

Exhibits 15.1 and 15.2 (data extracted from the website www.ourworldindata.org) provide a summary of international CO_2 emissions data for a limited number of countries. Exhibit 15.1 provides the data for "Production-Based-Emissions."

This is the most common way emissions data are presented. These data include emissions from all energy consumption activities inside a country's borders. The data in 15.2 are identified as "Consumption Based" and are "trade-adjusted" to take into account the energy consumed and emissions generated in one country to produce products that are exported to and "consumed" in another.

Exhibit 15.1 2017 Production-Based CO2 Emissions

Country	Annual CO_2 Emissions (Billions of Metric Tons)	%	Annual Per Capita Emissions (Metric Tons)	% Cumulative Emissions (1750-2017)
World	36.2	100	4.8	100
China	9.8	27.0	6.9	12.7
USA	5.3	15.0	16.2	25.0
EU	3.5	9.8	7.0	22.0
India	2.5	6.8	1.8	3.0
Japan	1.2	3.3	9.5	4.0
Saudi Arabia	0.64	1.8	19.0	0.9
Australia	0.41	1.1	17.0	1.1

Exhibit 15.2 2017 Consumption-Based CO_2 Emissions

Country	Annual CO_2 Emissions (Billions of Metric Tons)	%	Annual Per Capita Emissions (Metric Tons)
World	36.2	100	4.8
China	8.4	23.0	5.9
USA	5.7	16.0	17.4
EU	4.6	12.7	9.1
India	6.4	6.2	1.7
Japan	1.4	3.9	11.1
Saudi Arabia	0.63	1.7	18.7
Australia	0.37	1.0	15.5

For example, the emissions associated with manufacturing an air conditioner in China that's exported to, for example, the U.S., are added to the emissions totals for the importing country and subtracted from those of the exporter. The rationale for this method of emissions accounting is that the consumers in the importing country are the driving force for manufacturing the air conditioner. Absent their need (or desire) for an air conditioner, the emissions to produce the air conditioner would never have been generated. Consumption-based emissions-accounting attributes fewer total emissions to net <u>exporting</u> countries and greater total emissions to net <u>importing</u> countries.

Because global warming is effectively a cumulative phenomenon, Exhibit 15.1 also includes an estimate of the **cumulative production-based CO_2 emissions** by country from the mid-eighteenth century through 2017. (Sometimes readers may skip passed tables. You can't do that in this case. You need to study these tables and really understand what these numbers are saying.)

These data demonstrate lower cumulative contributions to atmospheric CO_2 from countries whose economies have expanded more recently such as China, India and Saudi Arabia, and conversely, larger cumulative responsibility for atmospheric CO_2 from countries whose energy-driven economies developed earlier.

Rich countries (e.g., the U.S.) and high population countries (e.g., China) tend to have high **total** emissions. Rich countries (e.g., the U.S.) and rich countries with lower population levels (e.g., Saudi Arabia, Australia) tend to have high **per capita emissions**. India is both poor and highly populated which makes its per capita CO_2 emissions quite low. Per capita CO_2 emissions in India are only about 10% of the per capita emissions in the U.S.! However, because the total population of India is so large, India is still a significant emitter of CO_2.

China is currently the world's largest total emitter (27% from China vs 15% from the U.S.), but the U.S. is the largest cumulative contributor to the total excess CO_2 concentration in the atmosphere (25% for the U.S. vs 12.7% for China).

There are a lot of possible takeaways from these tables. One that stands out is that the U.S. has been, and still is, a major contributor to global warming. From this, you can make a pretty strong ethical argument that the U.S. should be taking the lead to reduce GHG emissions and should be willing to take steps, even unilaterally, to move itself and the world along the path to a fossil-fuel-free future. That hasn't happened yet (and won't happen).

The data in these exhibits profoundly affect global warming negotiations between countries. The task of achieving international cooperation is further complicated by the direct impact of decarbonization efforts on the value of resources that some countries hold. Australia, India, and China, (and to a lesser extent, the U.S.), for example, all have vast coal reserves. If the use of coal is banned or rapidly reduced worldwide, these once-valuable national assets will become almost worthless. Constituencies within these countries are likely to resist the imposition of such limits or bans and in so doing, compromise international cooperation efforts.

International cooperation will also require some kind of global entity capable of bringing all the nations of the world together to reach a meaningful and enforceable emissions reduction agreement. **There is no such entity**. COP (Conference of the Parties) is a step in that direction, but it hasn't gotten the job done. The United Nations, for all of the good it can do around the world, hasn't even been able to stop the humanitarian catastrophe in the tiny country of Syria, so how can we reasonably expect that it would have any chance of wrangling all the nations together to end the use of fossil fuels? Even if such an entity existed, the whole endeavor would sound and feel like "world government" which, for a vocal minority in the U.S., is somehow more terrifying than wholesale environmental chaos.

Developing a coordinated, sustained, international commitment to combat global warming doesn't only run up against the real-world consideration that not all nations are contributing equally to the problem, it also has to contend with the fact that not all nations will suffer equal consequences. For both these reasons, a meaningful international agreement seems highly unlikely in the near term.

> *"Beating" climate change would require a coordinated, sustained international commitment that's unlikely to materialize.*

President Biden has appointed John Kerry to be the Special Presidential Envoy for Climate. For all the reasons previously discussed, it's fitting (and essential) that the U.S. be actively engaged with the international community to advance worldwide decarbonization policies, but Ambassador Kerry and his fellow negotiators have an impossible challenge. The delegations from all over the world will have to hammer out a plan that is far more aggressive than the Paris Accords (Miracle #1), then each of the delegations will have to return to their home countries to bring their citizens on-board to support the plan (Miracle #2) and finally each of the countries will have to implement their share of GHG emissions reductions (without cheating) (Miracle #3).

I would give Miracle #1 no more than a 10% chance of success, Miracle #2, a 5% chance of success (in Western democracies) and Miracle #3, a 5% chance. If these estimates are correct, the likelihood of a successful international effort to "beat" global warming is about 0.00025 (25 thousandths of one percent). Not great.

In regard to Miracle #1: there **will be** international agreements, but those agreements, like the Paris Accords, will be too timid. The delegates will pursue an incremental approach to "get the ball rolling," but each incremental action will run precious time off the clock, making the ultimate reduction targets that much stiffer and, consequently, that much less likely to be adopted.

Playing Games with the Numbers

President Biden has been doing what he can to inject a sense of urgency into the discussion of global warming. President Obama did the same. President Trump, on the other hand, did everything he could to prevent a meaningful international response to global warming.

The proponent of efforts to slow global warming, like President Biden and the leaders of most other nations, will do everything they can to maximize the appearance of progress to reduce emissions. Simultaneously, they'll also try to minimize the impact of emissions reductions on their own economies.

President Biden, for example, set a goal of reducing U.S. greenhouse gas emissions by 50% of what they were in 2005 by the year 2030. Why 2005? Because 2005 was the year when greenhouse gas emissions peaked in the U.S. Establishing the year of maximum emissions as the "baseline" against which future reductions will be measured makes it easier to claim larger percentage reductions. In April 2021 when the President announced his goal of reducing U.S. emissions by 50%, we were already 15% below that 2005 peak. Unfortunately, because of the pandemic and economic disruptions, we won't be able to draw any conclusions about how much real progress we're making (or not making) toward the President's goal until maybe 2024 at the earliest.

There is some logic to using the year of peak emissions as the base year against which reductions are measured. Specifically, it allows countries to begin undertaking serious emissions reduction efforts immediately even before any international reduction agreement is negotiated. In such an agreement, all nations presumably would be allowed to measure their emissions reductions from their peak emissions, taking credit for reductions prior to the agreement. Consequently, we need to be a bit reserved in how we celebrate emissions reduction progress. Because U.S. emissions in 2019 were already 15% below the 2005 peak, we were already roughly 30% of the way to meeting the President's emissions reduction target without having actually done anything new to reduce

> *Even if we achieve the President's 50% emissions reduction goal (we will not), U.S. per capita emissions would still only be on a par with those of European Union citizens today.*

emissions. And while all emissions reductions are good, even if we achieve the President's goal of reducing emissions by 50% by 2030, a substantial reduction, that achievement would only make U.S. per capita emissions roughly equivalent to that of European Union citizens . . . **today**. In other words, we have a long way to go to get off the "Worst Emitters" list.

Nation States Versus Humanity

Earlier in this text, I suggested that we (the collective "we") were many decades too late to "beat" global warming. By that I meant that we would only have had a viable chance of avoiding catastrophic global warming if we had started cutting back our use of fossil fuels before developing our unrestrained dependence on them.

However, I think you can argue that we actually "lost" any chance of "beating" global warming many thousands of years earlier. If international cooperation is required to "beat" global warming (which it is), it could be argued that we lost that battle when all of humanity began breaking up into tribes and then nation states. When we began thinking of those outside our local sphere as the "other," we were in competition with them, the "other," to stay alive and prosper. From that day forward, we were not "in" cooperation with those others. Collectively, we adopted an "us" vs. "them" worldview. For international cooperation to be truly successful, we would have to be living in a world where thinking of "us" meant thinking of all of humanity and not just our "tribe." When viewed this way, we would only have had the ability to come together to "beat" global warming today if humanity had embraced an entirely different playbook.

8 Billion AND Counting

Up until now, I've argued that beating climate change requires each of us to dramatically lower our carbon footprint, especially those of us in the rich, energy-intensive West. But another way for mankind to lower its collective carbon footprint would be if there were fewer of us making those footprints, i.e., some steady state population level that's substantially lower than it is today.

Unfortunately, the world's population isn't trending down. By November 2022, there will be 8,000,000,000 (8 billion) of us on the planet. Because the earth's population is skewed toward the young side, (i.e., more than 50% of all people on the planet are less than 30 years old) even if every couple around the world from this point until forever voluntarily agreed to have no more than two children each, the world's population would still swell from its current level to somewhere in the range of 11 billion people by the year 2100 (**one lifetime away**).

The Strain We Impose

There are just too many of us. In all likelihood, the world's population may have already exceeded the sustainable "carrying capacity" of the earth. Many have argued, and I agree, that the world is only able to support its existing population by relying on valuable, finite and in some cases, irreplaceable reservoirs of resources that are being draw down. Those resources are like money in the bank and the family of man has been living a lavish lifestyle by drawing down those savings. Eventually, that bank account will be emptied and that lavish lifestyle will come to an abrupt (and likely painful) end.

1. We're pumping our freshwater aquifers dry. Farmers and municipalities around the world are having to sink deeper and deeper wells to reach vital water supplies, much of which is used to grow food. Freshwater shortages will almost certainly be a forerunner to food shortages.
2. Conservative estimates are that we've depleted the world's natural fish stocks by more than 70% (some estimates say 90%). Fishermen around the world are working harder and harder for decreasing catches, a vital source of nutrition for a staggering fraction of the world's population. Even worse, we're altering the chemistry of the oceans in a way that can't support a vibrant ocean ecosystem.
3. We've consumed and are continuing to consume virtually all of the earth's easy-to-obtain fossil-fuel resources and are now going after the stuff that's much harder to get. We're fracturing gas-bearing rocks and squeezing oil out of tar sands.
4. We've dammed most of the world's major rivers disrupting the natural flow of water and silt.
5. We're exhausting the world's best farmlands. We've cut down (and are continuing to cut down) the world's old growth and ancient forests, in most cases, to gain access to only marginally productive farmland.
6. We've decimated most of the world's great wild animal herds. Species are going extinct at an accelerating rate.
7. We're melting the earth's great masses of ice which help stabilize our weather.

8. We're using up limited supplies of rare minerals, and,
9. (As relates to the topic of this book), we're using our own atmosphere as a garbage dump. We've spewed so much CO_2 and other GHGs into the atmosphere that we've fundamentally altered the heat balance of our entire planet.

We're depleting all of these "asset reservoirs" at an alarming rate to support the people we have on earth right now. What happens when these resource "bank accounts" are depleted?

Some of these natural reserves, once depleted, can't be restored. As water is being extracted from aquifers underneath Mexico City, for example, the porous rock structures in which the water is stored, have been collapsing. The land itself has been subsiding, destroying forever the ability of that stratum to store water. Some parts of Mexico City, the largest city in North America, are sinking at a rate of twenty inches per year due to water extraction from the aquifer under the city.

I'm a great believer in the restorative powers of nature. Given half a chance (and sometimes a little help), species on the brink of extinction have hung on and recovered. Plants and animals have returned to habitats that have been restored to their natural state. Ancient human settlements have been reclaimed by forests, jungles, and sand dunes.

But there are limits to nature's ability to heal the wounds inflicted on her when delicate balances are upset. Scientists believe that at one time Mars may have had water and an atmosphere, but both, if they were ever there, may have slipped away over time because the fragile balance that might have allowed them to be sustained weren't met.

Physical limits are real. A piece of metal can be subjected to moderate stresses that cause it to bend but then return to its original dimensions.

> **We're "stressing" the earth to the point where she may be permanently "deformed" or worse.**

Increasing those same stresses may cause the metal to be permanently deformed. Increase those stresses even further and the piece of metal may exceed its yield strength and break. We are, quite simply, stressing our earth to the point where she may be permanently "deformed" or worse.

If tapping into these vital reservoirs is what has allowed our population numbers to grow so dramatically and if we are now depleting these finite reservoirs, we may find that we can't sustainably support our earth's current population, much less one that's growing. If that's true, the question becomes: what population can the earth support? And, at what standard of living?

On one extreme is a standard of living and human achievement that's capable of landing a deep space probe on an asteroid to acquire and return to earth material samples from the primordial universe. On the other extreme are literally billions of people who, except for the cell phones they carry, are living lives not too dissimilar from the lives their ancestors lived 5000 years ago.

Energy and the Wealth of Nations

Look back at Exhibit 15.1 which included "per capita CO_2 emissions" by country. CO_2 emissions relate directly to energy consumption and energy consumption correlates (to some degree) with standard-of-living. Per capita emissions by the average American are almost four times those of the average citizen of the world and twice that of the average citizen of the EU.

Clearly there's plenty of room for us Americans to reign in our excessive and often wasteful lifestyles without reverting to the Stone Age. Perhaps a more prudent (and sustainable) standard of living might be more like that in the European Union: comfortable, modern, and advanced, for sure, but not as excessive as the American lifestyle. We Americans could easily reduce our per capita CO_2 emissions to the levels in the EU. The problem is, **if we reduced our per capita emission to EU levels and the rest of the world's population raised theirs to that level, the world's CO_2 emissions would at least double.**

There will always be inequities when comparing one group of people to another (both intra-nationally and internationally), but as the wealth and health of the average world citizen increases (a goal of individual governments and humanitarian organizations around the world), the consumption of goods and, unfortunately, the production of emissions of all kinds, including CO_2, will increase.

If we've already overshot the steady state level of population that the earth can sustain at a quality of life that most people might desire, nature has its own feedback mechanisms that have a way of bringing things back into equilibrium. Those mechanisms aren't pleasant. They include famine and disease. The inventiveness of man has partially thwarted some of nature's possible interventions to re-establish a balance. Antibiotics, vaccines, and modern medicine are counteracting some of the effects of disease. The "green revolution" of the 1950s, '60s and '70s (high yield crop varieties, intensive use of fertilizers, pesticides, and other modern farming techniques) has, for the time being, counteracted some of the effects of famine. Advanced extraction technologies (such as directional drilling, fracking, and tar sands processing) have removed some of the world's energy constraints, albeit to the detriment of the planet.

All these human achievements raised the life expectancy for billions of people and allowed populations to grow. They have also removed natural population constraints and have in effect allowed us to wind the spring in mankind's clock tighter and tighter and tighter. We all know there are limits to how tightly we can wind that spring.

Sustaining the population we already have will become increasingly difficult as the reservoirs of natural assets are depleted. As the supply of "low-hanging fruit" runs out, it will take even more energy, not less, to sustain the already energy-intensive quality of life to which we (in developed countries) have become accustomed. Delivering those same benefits to a growing population will be even that much harder – harder on us and harder on our environment.

As medical science has improved, as populations have become richer, and as women have become better educated, worldwide fertility rates have been dropping. This is a good trend in terms of slowing the world's population growth, but a decreasing fertility rate is typically a "lagging indicator." First the standard of living (food, water, shelter, security, medical care, physical things, etc.) has to improve, and then, only after at least one generation has enjoyed that improved standard of living, will fertility rates begin to decline. These "wealth-based" natural reductions in fertility rates have been on-going in parallel with lots of national and international humanitarian programs that seek to reduce birth rates. Many of these efforts have been effective, but in spite of these successes, the world population is still increasing. This year alone, the world's population will grow by another eighty-three million people. **Since just 2010, the world's population has increased by an astounding one billion people!**

China's one-child policy established in 1979, is estimated to have reduced China's current population by 300 million people compared to what it would have been without that policy. In other words, China is almost "one entire U.S. population" less populous than it would have been without that policy. China abandoned its one-child policy and established a two-child policy in 2016, but many Chinese couples are still choosing to limit the size of their families.

Impacts Beyond Our Numbers

One last point about population. Eight billion of anything is a LOT of . . . anything. (Aren't you glad you bought this book for these kinds of insightful observations.)

Anyway, eight billion people is a LOT of people, especially when you think about the fact that our species, homo sapiens, only first emerged from the genetic lottery some 200,000 years ago and only first settled down in tiny agricultural communities scattered all over the planet in the last 10,000 years.

As we got smarter and learned more about our world, we multiplied and began to manipulate our surroundings. It's truly staggering to think about how much knowledge we've gained in the last 1000 years or even just the last one hundred years and how much easier and more certain our lives are today compared to the experience of our grandparents or even our parents. (My grandparents, for example, were born into a world without antibiotics. Well into the 20th century, before penicillin was widely available, patients with common ear infections were routinely admitted to hospitals for treatment.)

Mankind has made enormous progress, but we've also had an enormous impact on our earth. In the short time we've been the dominant animal life species, we've altered our environment in ways that impact all living things, plant and animal.

And yet, in contrast to the impact we've had, **we are but a minuscule fraction of a minuscule fraction of this earth.**

The **average** human being is less than 6 feet tall and for the sake of argument weighs less than around 200 pounds. That means that each of us would easily fit in a space approximately six feet long and one foot on each side; each of us would occupy a volume of well less than around 6 cubic feet.

If all of humanity, all 8 billion of us, were to lay down (like logs at a lumber mill) in New York's Central Park, that stack of humanity would reach just 1,308 feet into the sky – more than 450 feet <u>shorter</u> than the Freedom Tower! **Every single one of us** -- all the doctors, all the lawyers, all the drill rig operators, all the farmers, all the bus drivers, all the bloggers, all the grocery store clerks, all the teachers, all the researchers, all the Uber drivers, all the grade school kids, all the seniors – EVERYBODY FROM EVERY COUNTRY ON EARTH -- every last one of us would fit in that tiny volume. Don't believe it? I didn't . . . until I ran the numbers.

For perspective, that volume to humanity would cover the entire island of Manhattan (22.8 square miles) up to a height of just 77 feet. Or the Greek island of Mykonos up to a height of 53 feet. Or the island of Hong Kong up to a height of 56 feet. Or the political "island" of Washington D.C. up to a height of 26 feet.

> *How could such a small "volume" of (arguably) intelligent living matter have such a significant impact on an entire planet, know that it was destroying its only home, and still* <u>*do almost nothing about it?*</u>

How could this be? How could such a small "volume" of living matter have such an enormous impact on an entire planet? And even **more incredibly,** how could such a small volume of *(arguably)* intelligent living matter know that it was destroying its only home and still **do almost nothing about it?**

A Mini-Recap

Because CO$_2$, global temperature and weather don't respect country borders, international cooperation is a requirement for us to have any hope of even just slowing global warming. Unfortunately, there's no evidence that the necessary global cooperation, commitment, and follow-through can or will be achieved.

Take a look at Exhibit 15.3. In this exhibit I've recreated the plot of CO$_2$ concentration in the atmosphere over the last 60 years (Exhibit 3.1) and over-laid onto that curve some notable events related to global warming. Sadly, this combined plot speaks **volumes.** In spite of greater knowledge about the phenomenon, greater worldwide awareness of the problem, progress in deploying renewable resources, and international attempts to limit greenhouse gas emissions, **the buildup of CO$_2$ in the atmosphere has not only <u>not</u> gone down, the concentration of GHGs as gone up and the rate at which its going up has actually increased** (currently 3 ppm/year). We're not only <u>not</u> winning this battle, we're losing it by a larger margin by the hour.

Exhibit 15.3
Timeline of INACTION

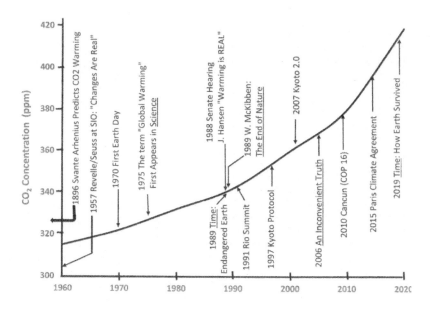

The good news is that carbon-free resources are finally beginning to be deployed around the world in numbers that could materially slow the buildup of CO_2 in the atmosphere and its absorption by the oceans. However, because of COVID-19 disruptions to the world economy and the war in Ukraine, it may take several years for worldwide economic activity to come back to "normal" so that we'll know with any confidence how much of a difference our efforts toward decarbonization are making.

Progress toward decarbonization can be measured in atmospheric GHG concentrations, but the processes leading to those changes are complex and multi-dimensional. Energy generation is impacting our climate and our weather which are in turn negatively impacting worldwide food production. Ocean acidification, ocean warming, and unsustainable fishing practices are continuing to deplete fish populations. Catches of wild fish are decreasing which is putting more pressure on land-farming to make up the nutritional deficit. As food shortages develop, crops such as corn, that are currently being diverted to the production of ethanol, should have a higher value as part of the human/animal food chain. Ethanol production plants will have to use other less energy-dense feedstocks or if the ethanol plants are shutdown, petroleum usage will increase.

Food shortages will also put increased pressure on rainforests. Increases in irrigated farming have caused the extraction of massive quantities of freshwater from underground aquifers, forcing farmers (and cities) to drill ever deeper wells and invest in infrastructure to move freshwater from one area to another. Warmer temperatures are increasing the number of degree-cooling days which will show up as increased energy consumption for space conditioning. All of these dynamics (and many others) are interrelated.

Our predicament is this: with reserves of natural resources being drawn down at unsustainable rates, we'll have to expend more energy, not less, to fulfill even the basic needs of food, water and shelter for the world's growing population. The

> *We will have to expend more energy, not less, to fulfill even the basic needs of food, water and shelter for the world's growing population.*

greater energy intensity associated with supplying basic human needs, the growing global population, and efforts to raise the standard of living for the less-well-off will offset many of the emissions reductions we might hope to realize by deploying carbon-free energy resources (solar and wind). In other words, it's going to be hard to decrease the rate at which CO$_2$ is building up in the atmosphere. It's going to be hard to go from 3 ppm per year increases down to even 1.5 ppm per year increases, let alone zero (or negative).

At the conclusion of the Paris Climate Conference, delegates agreed on at least one thing: **they needed to do better next time.**

> *It's going to be hard to bend the curve of increasing CO$_2$ concentrations down from 3 ppm/year.*

[Breaking News: This just in. Glasgow, Scotland, COP26, November 2021. **They didn't.**]

16 TECHNOLOGICAL FIXES

**"Build a better mousetrap and the world
will beat a path to your door."**

**(Attributed to) Ralph Waldo Emerson
Late 19th Century**

We could end most of the excess man-made emissions of CO_2 (and other GHGs) if we displaced fossil fuels from electricity generation, transportation, and industrial processes, ended direct combustion in residential and commercial applications and re-imagined agricultural production.

We've made progress and will continue to make progress on all of these fronts, but I've argued that ending our reliance on fossil fuels in all these areas will be glacially slow and monumentally expensive if we limit ourselves to solar and wind generation. I've also argued that the transition to a clean transportation sector is a **new** electrical demand and is dependent on a much-expanded clean electricity grid. I've explained that zero-tailpipe-emission EVs are better than most of our current fleet of internal-combustion-engine powered cars, but that even EVs will continue to be a significant source of emissions until the grid from which they are being recharged is carbon-free. I've also argued that decarbonizing the industrial, commercial, and residential sectors will also be an entirely **new** electrical load and that the conversion of this element of our fossil fuel usage will increase CO_2 emissions until the energy sources powering them are also carbon-free.

If, as I argue, the carbon-free electric generation technologies available to us to displace fossil fuels are painfully few, it would appear as though we **need** technology breakthroughs (in a number of areas) to save our bacon. Speaking of bacon, let's look at agriculture.

What's For Dinner?

The agricultural sector (the production, processing, and distribution of food products) is essentially an industrial enterprise where the profit motive has already driven corporations to exploit the economies of scale and energy conservation strategies to minimize costs. Further emissions reductions in this sector will materialize as the electric grid and transportation sector reductions are achieved. There is, however, one aspect of the agricultural sector where consumers could reduce food-production-related GHG emissions. We could change our diets.

Eating meat is a personal choice (for most people), but as explained in Chapter 12, producing beef consumes a lot of water and grain with the additional environmental insult of producing methane. While most people can't simply unplug from the electrical grid, they could, if they wanted to, dramatically reduce the quantity of meat they consume. There are, however, powerful lobbies, advertising interests and personal biases that will try to deter us from adopting a more plant-based diet.

Case in point from the category "whose ox is being gored:" TIME Magazine reported that the U.S. Cattlemen's Association has been lobbying for a federal law requiring manufacturers to put the word "imitation" on labels of plant-based and cell-cultured meat replacements. Dozens of states have already or are now pursuing labeling for this class of replacement protein to include the words "imitation meat" or something similar. Presumably, the Cattlemen's Association believes that burdening the plant-based meat substitutes with the label "imitation" will discourage their acceptance in the marketplace. This is a perfect example of **powerful forces** attempting to perpetuate the status quo for their own profitability. [Turnabout is fair play: perhaps "imitation" meat producers should lobby their state legislatures to require packages of meat to have the label, "slaughtered animal carcass."]

The difference, however, between dietary choices and the other kinds of changes I've discussed relating to the electric grid and the transportation sectors is that changing to a more vegetable-based diet doesn't require a significant financial investment by the consumer. You don't have to buy a Tesla or put solar panels on your roof to change your diet. The cost premium for plant-based protein substitutes will decrease over time making it an easier choice for consumers. In any case, dietary choices to reduce CO_2 emissions can take place gradually, or overnight.

The reason for raising the issue of meat consumption in this chapter on technological fixes is that a lot of work is being done to develop and expand the production of plant-based and cell-cultured meat substitutes. The extent to which these substitutes can help reduce GHG emissions will play out over the next few decades. In the meantime, meat consumption in the U.S. has not dropped in recent years. In fact, prior to COVID-19, it had reached an all-time high. Globally, meat consumption has increased year over year for the last ten years at a rate twice that of population growth. This is not a trend consistent with "beating" global warming.

Left at the Altar

In Chapter 7 I discussed a proven, non-intermittent technology with centuries of fuel reserves that doesn't emit any CO_2, namely nuclear power. I mention it again here before discussing the areas where we need breakthroughs in energy production technologies as a reminder that we already have a carbon-free generating technology that could be pressed into service if we could only develop a consensus to do so.

Until we develop that consensus, I've offered my opinion that this clean power source will not make a comeback (certainly not in the U.S.) until we've tried virtually every alternative. I've also offered the opinion that two things need to happen before **new nuclear power plants** can make a dent in U.S. CO_2 emissions. First, nuclear vendors, constructors, utilities, and regulators have to successfully build and operate one or two of the next-generation nuclear plants that are smaller, standardized, passively safe, utilize factory-fabricated subsystems, and can be licensed and built in under five years. **And,** if that weren't enough, second, we will have to make a permanent, high-level, nuclear waste repository operational. Even if those two things happen, the deployment of a meaningful number of new nuclear power plants is not going to happen in the next thirty years. It **will** happen eventually, but not until the public is convinced that there are no alternatives – and that probably won't happen until the 2050 or 2075 time frame.

Five Tech Fixes and Their Fatal Flaws

Without nuclear power, where does that leave us? We have to come up with one or more of the following:

1. Other carbon-free electric generating options that are at least as cheap as solar panels and ideally are not intermittent in nature,
2. Entirely new technologies to synthesize storable, mobile fuels from feedstocks of CO_2 and water,
3. New, inexpensive, efficient, and safe electrochemical battery options whose manufacturing can be ramped up in a matter of years,
4. New, inexpensive technologies to extract CO_2 from combustion product waste streams (or the free atmosphere) and a way to store unfathomably massive quantities of that CO_2 underground safely and cheaply forever, and
5. A way to minimize the amount of solar radiation impinging upon and ultimately being absorbed in the earth's atmosphere.

To address the first category of technology solutions, i.e., "other" green electric generating technologies, I need to revisit a point I made in Chapter 13, Economics.

In Search of the Holy Grail

Very smart people have been searching for the holy grail of energy production for at least the past seventy years. Initially, the motivation wasn't climate change. The fear was that we'd run out of cheap fossil fuels. Anticipating a shortage (which hasn't materialized) and much higher costs (which have), researchers around the world have worked on every conceivable form of energy generation. Civilian nuclear power production was one of the fruits of that labor. Solar panels also emerged from that effort and were progressively improved upon for the space program, where efficiency and weight were paramount. Windmills weren't new, but modern, efficient, large, durable, electric-generating wind turbines have evolved over the past forty years.

With new nuclear power plants effectively halted (in the U.S.) wind and solar technologies emerged as "winners" in the free-market competition for new, clean energy sources. Dozens of other technologies were conceived, prototyped, and tested, but none could beat the two primary, green technologies we're still deploying today. If there were a better mousetrap, I assure you: we'd be building them by the thousands of megawatts and their developers would eclipse the wealth of Musk, Gates and Bezos combined.

The sad hard truth is that with the exception of nuclear fission, these clean, efficient, cost-competitive generation options just aren't out there. Mind you, I'm not saying that cost and efficiency improvements aren't to be had. They are. What I am saying is that the prospect of discovering or inventing a magic bullet that's going to completely transform the world's energy framework in the next thirty years is highly unlikely.

> *The sad hard truth is that with the exception of nuclear fission, clean, efficient, non-intermittent, cost-competitive generation options just aren't out there.*

A Promise Unfulfilled

Fusion power is often identified as the holy grail of energy technologies. With apologies to my former classmates who followed that technology path, viable fusion power plants are forty years in the future – they **always have been, and more importantly, always will be**

Unlike the nuclear power plants of today that produce energy by splitting high atomic number nuclei (a process called fission), a fusion nuclear power plant would produce energy by fusing two low atomic number nuclei into a single heavier nucleus. The two most prominent fusion reactions under consideration for use in a fusion reactor involve isotopes of hydrogen: D-D (deuterium-deuterium) and D-T (deuterium-tritium). Both of these reactions release extremely high-energy neutrons when the two nuclei fuse.

Igniting a fusion reaction is hard. Because the nuclei being fused are positively charged, they naturally repel one another. In order for the fusion reaction to take place, the two nuclei have to be forced "close enough" to each other for the reaction to proceed. The mass at the center of our solar system is so great that the force of gravity accomplishes this feat in the fusion reactor we affectionately call the sun. Gravity crushes the positively charged hydrogen nuclei so close together that they fuse to form the nucleus of a helium atom and release enormous quantities of energy in the process.

The technical challenge of getting hydrogen nuclei to fuse on Earth is to build a machine that will compress a plasma consisting of the nuclei of the fuel atoms to a high enough density for a long enough duration to allow the fusion reaction to occur – and, in the process, produce more usable energy than the machine consumes.

Building a successful fusion device may well turn out to be the most challenging physics experiment ever undertaken. The superconducting magnets that generate the magnetic fields to confine the charged-particle plasma require temperatures close to absolute zero while the temperature of the plasma will be about 150,000,000° C.

A large, advanced fusion device is currently under construction in southern France. The project is called ITER, the International Thermonuclear Experimental Reactor. An international consortium which includes the U.S. is funding this $22 billion project. Construction began in 2013, but with delays, design changes and cost overruns, ITER will not be finished before 2025. If and when ITER is finally operational, the facility will not start testing the most promising fusion reaction (D-T) until 2035.

The most important word in the ITER facility's name is the word "Experimental." The plasma it produces may possibly achieve "energy break-even" (i.e., where the energy output of the plasma exceeds the energy needed to ignite the fuel), but it's important to remember that ITER is, as its name makes clear, an experiment, a large and expensive **physics and materials experiment**. It doesn't even include a power plant because at this stage of fusion development, a power plant would be an unnecessary and costly distraction. All the energy consumed to ignite the plasma and all the energy produced in the plasma itself (if successful) will be vented to the atmosphere.

In 1983 Dr. Lawrence M. Lidsky, the MIT professor who introduced two decades of nuclear engineering graduate students (including myself) to controlled nuclear fusion, published a paper in the MIT Technology Review titled "The Trouble with Fusion." Dr. Lidsky devoted his entire career to fusion energy research which led him to conclude that a D-T fusion reactor could never be commercially viable. One objective of ITER is to test one of Dr. Lidsky's many concerns, i.e., to see if the reactor's inner wall can withstand the bombardment of extremely high energy neutrons. Under such a bombardment many of the reactor's internal components will become so radioactive that hands-on maintenance would be impossible. (There are fusion reactions other than D-T that don't produce as much damaging radiation, but those reactions are more difficult to ignite).

To the best of my knowledge, most if not all, of the and commercial concerns Dr. Lidsky identified are still open questi͟o͟n͟

ITER, a "donut-shaped" fusion reactor known as a tokamak, is just one fusion concept. There are several others in the conceptual stage. One of them is referred to as a "compact fusion reactor." It too is a tokamak reactor, but physically much smaller than ITER because it takes advantage of more recent developments in smaller, high-temperature superconducting magnets to generate the magnetic fields required to confine the energy-producing plasma. The promoters of the concept are raising money and proposing a development schedule that's much shorter than ITER. If, however, any of the compact fusion concepts were sufficiently advanced to the point where a universally held belief existed among plasma physicists that compact fusion devices (or any of the other fusion concepts) had a significantly better chance of technical success than ITER, I assume funds would be redirected to these more promising technologies. That hasn't happened.

Even more importantly, if I'm wrong about the eventual technical success of any fusion concept, a fusion power plant is unlikely to ever be commercially viable. Long before any fusion power plant will be grid-ready, other less-costly, less-risky, non-CO_2-producing fission power plants with all of the same important long-term benefits ascribed to fusion will have been operational for decades on the very distant electric grid of the future. (This is precisely the same obstacle that alternative renewable technologies face today – the need to compete with proven, less expensive technologies, namely wind and solar power generation.)

Even if every technical challenge Dr. Lidsky identified is eventually overcome (as well as some he may not have foreseen), the future fusion power plant, he concluded, "is in danger of joining the ranks of other technical 'triumphs' such as the Zeppelin [and] supersonic transport . . ."

Progress in plasma physics experiments over the past fifty years has been incredible, but I can say without hesitation, nobody

> *Nobody alive today will ever enjoy a single kilowatt-hour of energy from a fusion reactor. It's often said that prudent people never say "never." In the case of fusion power, I'm not worried.*

alive today will ever enjoy a single kilowatt-hour of energy from a fusion reactor. (It's often said that prudent people never say "never." In the case of fusion power, I'm not worried). If I were King, I'd terminate funding for fusion reactors and redirect that money to nearer term, more proven technologies.

Incidentally, if I were King, I would not only terminate fusion power research, but other research efforts would also be on the chopping block. Unrelated to energy issues, I'd terminate spending for all **manned** space travel beyond Earth orbit including a return to the moon and a mission to Mars. I'd also ax development of a new supersonic jetliner. I'm not a Luddite, and I'm not opposed to funding basic

> *Forgo* **manned** *Mars missions and focus on Mother Earth. Mars will still be there in 2150, virtually unchanged from how it is today. Earth, on the other hand, is likely to be a decidedly less hospitable place by then.*

research, but as the effects of global warming begin to consume more and more resources, the public will demand that funding be directed to projects with more significant near-term benefits for life on Earth. And anyway, if we "weather" the global warming crisis here on Earth in the next 100 years, Mars will still be out there in 2150, virtually unchanged from how it is today. Earth, on the other hand, is likely to be a radically different and decidedly less hospitable place than it is today.

New Mobile Fuels

The second category of technological fixes involves developing new methods of synthesizing fuels from water and carbon dioxide. Researchers are attempting to develop advanced techniques to use sunlight to break down water into its component parts (hydrogen and oxygen) using a variety of different catalysts. Because this process mimics a portion of what takes place in plant photosynthesis, this research is sometimes referred to as the "artificial leaf." The hydrogen generated in the process could be used as a fuel (without producing any GHGs) or further processed to create methanol, an industrial feedstock or liquid fuel.

The artificial leaf and other processes will generate hopeful headlines in coming years, but for now, they are laboratory-scale experiments. Because we're on the leading edge of severe global warming consequences **right now**, only those technologies that are already in the field or in the advanced demonstration phase today will have any chance of helping the world actually combat global warming. As promising as some of this research is, these technologies, even if successful, are, like fusion and advanced nuclear fission reactors, likely to arrive too late and can't be deployed quickly enough to make a significant difference in curbing CO_2 emissions in the near term.

Energy in a Bottle

The third category of technological fixes involves cheap, efficient energy storage.

A lot of research and development dollars are being directed toward battery system concepts. A few new battery technologies with large-scale potential are likely to emerge from this lab work. Some of these will be perfected and perhaps a few may ultimately be commercialized. Some will be used in EVs, some will be used in commercial, warehouse-sized energy storage centers, and some will be used in home energy storage applications. Unfortunately, however, I don't believe we'll see dramatic cost reductions and/or efficiency improvements from any of these new batteries. The most promising (i.e., cost-competitive) combinations of chemical constituents have already been investigated. Any new battery type will have to compete with the proven technologies that we already have and will most likely suffer the cost disadvantage of requiring rare and expensive materials. Solid state batteries show a great deal of promise, but the performance of these batteries appear to degrade after a number of charge-discharge cycles.

After deploying a substantially quantity of carbon-free resources on the current electrical grid and the expanded grid to serve new electrical demands (EVs, electrified heavy rail, process heat, etc.), the most urgent need (in terms of combating climate change) will be for large, stationary, bulk energy storage batteries. Low initial cost, a long cycle life, low maintenance requirements and safety are likely to be the most critical design considerations. Weight, a critical design consideration for EVs, will be of little concern in fixed installations.

In short, I think there will be improvements in energy storage technologies, especially in manufacturing and mass production to bring costs down, but I strongly doubt that any new battery concept will emerge that provides even a factor of two cost improvement over what's currently available and that level of improvement won't be enough to make battery energy storage an inexpensive replacement for fossil fuel electric generation at times when solar and wind power are unavailable.

Capturing CO_2

The fourth category of technological fixes goes by the names "carbon capture and storage" (CCS) or "carbon dioxide removal" (CDR). Carbon capture is already being done commercially when the captured CO_2 can be sold as a component for use in industrial and chemical processes (e.g., making soft drinks, manufacturing biofuels, etc.), but those quantities are minuscule compared to the total amount of CO_2 being emitted annually and already residing in the atmosphere. CO_2 is also being sold to some greenhouses to boost plant growth, but again, the quantities are minuscule. In any case, most, if not all of the CO_2 captured for use in these commercial applications, is released back into the atmosphere eventually, so there's no net reduction in atmospheric CO_2.

The idea behind wholesale CCS is simple: since we have too much CO_2 in the atmosphere, we need to take some of it out (that's the "capture" part) and then put it somewhere (that's the "storage" part). Once extracted from the atmosphere, the CO_2 gas would be put in a stable chemical form or compressed into a liquid (above around 75 psi) for storage in a secure, underground, monitored facility.

Many large-scale CCS concepts are technically feasible, but just don't make any sense.

First, consider this analogy. Let's say you're a doctor and you have a patient bleeding out on the operating table. Because we've all seen **too many** hospital dramas on TV, we know that the first course of action to save the patient is to stop the bleeding, i.e., stop the progression of the immediate condition that's threatening the patient. In the case of global warming, "stop the bleeding" means stop dumping CO_2 into the atmosphere. Stop burning fossil fuels. Shouldn't that be the first course of action to save patient Earth? Wouldn't that make more sense than expending a lot of money to pull a ton of CO_2 out of the atmosphere at one industrial site while a power plant across town is spewing out 5 tons of CO_2 per hour?

Carbon dioxide gas might be dilute (i.e., in a low concentration in the air and mixed with other gases) or more concentrated. The atmosphere, which contains roughly 0.042% CO_2, (420 ppm), is an example of CO_2 in a dilute form. The gas streams from processes that produces CO_2, such as the effluent from a coal- or gas-burning power plant or a cement manufacturing plant are examples of concentrated CO_2 gas streams. If your goal is to extract as much CO_2 as you can by processing the least volume of gas, it makes sense to focus on the concentrated CO_2-rich gas streams first.

Coal-burning power plants are an obvious choice for this process because their effluent contains a higher concentration of CO_2 than even a natural-gas-burning power plant. In addition, those who support the burning of coal without restrictions are anxious to show that coal can be made clean. (It can't be.) But still, there have been CCS projects to demonstrate the technology. My tax money and yours paid for these demonstrations. Tell me if this is how you would have wanted **your** tax money spent.

First, coal is dug out of the ground. Then it's cleaned (usually with water), transported to a power plant (usually by diesel-powered trains) and unloaded into a huge pile. When the power plant needs fuel, conveyor belts transport coal to a processing plant where it's pulverized into a fine dust before being injected into the boiler, where it burns and releases energy.

In state-of-the-art coal power plants, most of the energy released in the combustion process is transferred to water flowing through tubes inside the combustion chamber. A small fraction of the combustion energy remains in the hot combustion gases and is discharged out the smokestack into the atmosphere. (Most modern plants treat this effluent gas stream to remove a large fraction of the most onerous pollutants such as sulfur dioxide and fine particulates, but that's another discussion).

In the power plant, water flowing in tubes lining the walls of the combustion chamber is converted to steam at high pressure. The steam flows through large pipes to the turbine-generator. As the steam cascades through the blades of the turbine, losing temperature and pressure along the way, it forces the shaft of the turbine to rotate. The turbine is connected to an electric generator and, voilà, electricity.

At the exit of the turbine, the steam is now at a lower temperature and pressure, but still has to be condensed back into water and pumped in a closed loop back to the tubes lining the walls of the combustion chamber. This final cooling process occurs in the condenser where the ultimate heat-sink is typically the ocean, a lake, a river, or the air.

Just over a third of the chemical energy in the coal dug out of the ground is converted into electrical energy. **Two-thirds is discharged as rejected heat** (which I discussed in Chapter 6). Higher conversion efficiencies for coal are possible by going to higher temperatures and pressures or using alternative conversion processes, but all of them cost so much more that using them would increase the cost of electricity, not decrease it. While the example I am using here is a coal-burning power plant, all thermal power plants regardless of the heat source (coal, gas, nuclear, geothermal, thermal solar, wood chips) use the same energy conversion process (the Rankine cycle) to generate electricity. Suffice it to say that physical laws, material limitations and economics all conspire to limit the efficiency of most thermal power plants to somewhere in the mid-thirty percent range. (Natural gas burning combined cycle power plants that utilize a gas turbine (a jet engine) in conjunction with a steam turbine can achieve fuel efficiencies of around 60%). Most new natural-gas-burning generating capacity utilize a combined cycle configuration; most of the older, existing natural-gas-burning .generating capacity utilizes just the lower efficiency steam turbine cycle.

The above describes how electricity is produced in all coal-burning power plants. The part of the plant that's different in a power plant equipped with a CCS system involves the processing of the effluent from the plant's combustion chamber before it's released from the plant's smokestack.

The CCS system extracts the CO_2 from this hot gaseous effluent. The first step in that process is to cool the combustion gases and scrub out fine particulate matter. These cooler, cleaner gases are then pumped into an absorption column containing a solvent that preferentially absorbs CO_2. The solvent, now laden with CO_2, is pumped out of the absorption column to a chamber where it's reheated to drive off the CO_2. The solvent is then cooled and recycled back to the absorption column to begin the process all over again.

The stream of CO_2 gas released from the solvent is then cooled again and ultimately compressed into a liquid which can be injected into some sort of underground chamber such as a depleted oil reservoir, salt cavern or abandoned coal bed. There are other technologies to capture CO_2, but this one illustrates the basic process.

OK. So, this is what we have. First, we build a power plant and then, right next to it, we build a gas processing plant. Then you dig carbon out of the ground to burn in your power plant with an efficiency of approximately 34%. Then you divert a portion of the electric energy generated in your power plant to your gas-processing plant where you use it to repeatedly cool and then reheat combustion gases and solvent, and to move gases and liquids around inside the gas processing plant. Then, you use still more energy from the power plant to compress the CO_2 gas into a liquid and finally you use even more energy to pump the CO_2 liquid deep into the earth.

Does it work? Yes. CO_2 that would have been released into the atmosphere is instead underground. Is it expensive? What do you think?

After making the capital investment to build the gas processing plant, incurring the expenses to operate and maintain it and consuming a significant fraction of the power plant's electrical output to run the CCS system, the end result is that the remaining electric energy to sell to off-site customers costs about double what it would cost without the CCS system.

If the price of energy from a fossil-fuel power plant augmented with a CCS plant is twice as high as it would have been from the fossil fuel plant alone, that's a little like extracting the fossil fuel from the earth with the expectation of getting just 18% of the total energy content of the fuel for anything useful.

From my perspective, the more sensible alternative to carbon capture and storage (CCS) is to leave the carbon in the ground where it's been for millions of years.

I've described just one process for capturing CO_2 from a carbon-rich gaseous effluent stream, and I'm clearly not a fan. There are other high-tech processes to extract CO_2 from carbon-rich gas streams and ambient air. One concept uses fuel cells in combination with calcium carbonate (limestone) kilns, to generate electricity, produce cement and pull CO_2 out of the effluent gas stream. The fuel cells produce heat for the limestone kilns to drive CO_2 out of the limestone, but the energy source for the fuel cell is natural gas – which means that the process, end to end, would still produce CO_2.

> *By far, the more sensible alternative to carbon capture and storage (CCS) is to simply leave the carbon in the ground where it's been for millions of years.*

Research is also being conducted on ways to extract CO_2 from ambient air as it flows over specialized membranes. There are also low-tech approaches to promote the uptake of CO_2 by seaweed. All of the CCS concepts, whether conceptual, under evaluation or proven, face additional obstacles.

In addition to the expense of building and operating a CCS plant there are some practical physical constraints. If we really pursued CCS on a scale that would substantially draw down the concentration of CO_2 in the atmosphere (especially if we simultaneously continue burning ever more fossil fuel), where would we put all that CO_2?

There's also a question of safety in facilities that would store liquefied CO_2. At atmospheric pressure, CO_2 is a heavier-than-air gas. If CO_2 gas were to leak out of its underground bunker, it would settle close to the earth's surface, where, as it happens, the world's people and animals live. Surface winds would eventually mix and disperse the gas, but there are recorded cases of natural CO_2 releases in the vicinity of volcanoes that have remained intact at ground level long enough and in concentrations high enough to suffocate humans and other animals. A mixture of 10% CO_2 with 90% N_2 and O_2 can be lethal. The maximum permissible occupational exposure is one twentieth that concentration, 0.5% CO_2.

CCS has another hurdle that relates to the discussion in Chapter 15 on international cooperation. Specifically, the issue is this: who's going to pay for this? Because wind and water currents don't respect country borders, CCS initiatives raise all the same issues as the international carbon tax. If some country fires up a massive number of CCS facilities, who will reimburse them for their global good deed and by how much? Initially, the product of CCS enterprises will be carbon credits. These carbon credits will be sold (if cap and trade programs persist) to cities, countries, and companies to offset CO_2 emissions they can't or don't want to stop making. In other words, the captured carbon will not actually reduce the amount of CO_2 in the atmosphere, it will only help slow the rate at which the atmosphere is getting worse.

The last point I'd like to make in regard to CCS is the same one I made in explaining why new nuclear power plants aren't going to save the planet. Both approaches could slow down the rate at which CO_2 builds up in the atmosphere, but only if they're brought online quickly, in substantial numbers (and, in the case of CCS), don't themselves represent a significant additional electrical load). To really draw down the concentration of CO_2 in the atmosphere, first we have to stop burning fossil fuels and then we'd have to deploy untold numbers of CCS plants in the next three decades. I don't see either of those things happening. CCS will not allow us to "beat" global warming in the near term.

[*Input] I'm certain I haven't done justice to many CCS concepts, so I invite knowledgeable individuals to submit additional information for the second edition. In your article, please be clear about the status of the development, potential for scale-up to commercial operations, operating equipment, material flows and cost estimates. In other words, provide at least enough detail to know if the technology is deployable on a large scale by at least 2050 to credibly draw down CO_2 concentrations. [*Input – End]

Geoengineering – Here Comes the Sun (Shade)

The fifth area of technological fixes to decrease global warming is to attack the heating source directly, i.e., to reduce the amount of the sun's radiation impinging on the earth's atmosphere. These global warming countermeasures are called geoengineering or solar radiation management (SRM). These are man-made efforts to alter the physical environment to fix a man-made problem. (Is my lack of enthusiasm for what I'm about to discuss obvious?)

Because we don't have a dimmer switch for our sun, some researchers have proposed putting up a sunshade of sorts to block a fraction of the sun's rays from being absorbed in the earth's atmosphere. Various forms of sunshades have been proposed. One idea is to launch giant, foldable mirrors (or untold numbers of smaller mirrors) into space to reflect a small percentage of the sun's radiation before it strikes the earth.

Another approach would be to increase the quantity of high clouds, especially over the oceans, where the clouds would reflect more of the incident radiation back into space before it has a chance to be absorbed in the darker ocean below. The terrestrial equivalent of increasing the cloud cover is to increase the reflectivity of the earth (i.e., increase its surface albedo) which could be accomplished by, among other things, painting roofs and other large, upward-facing surfaces white.

And yet another idea is to inject aerosols of sulfur compounds into stratosphere, much the way a volcano does during a large and energetic eruption. Sulfur dioxide gas can dissolve in water vapor and combine chemically with oxygen to make sulfuric acid, H_2SO_4. These sulfur compounds would create a haze in the stratosphere that would reflect sunlight out into space before it could penetrate down into our lower atmosphere and be absorbed.

All of these concepts may be technically feasible, so the questions that arise relate to:

1. Effectiveness,
2. Cost, and
3. Possible harmful side effects.

[*Input] As with the prior section on CCS, this is not my area of expertise, so again, I'd invite anyone with particular expertise in this area to make a submission at BlueOasisNoMore.com for inclusion in the second edition. [*Input – End]

One drawback shared by **all** of these SRM concepts is that they do nothing to lower the concentration of CO_2 in the atmosphere. Instead, they seek to reduce the net energy being absorbed by the atmosphere and the land below by reducing the amount of radiation available to be absorbed. If the only consequence of added CO_2 in the atmosphere was global warming, these SRM concepts might be more appealing, but the excess CO_2 in the atmosphere causes other damage. Recall that our oceans absorb between a third and a half of all atmospheric CO_2 causing ocean acidification. Scientists have already documented coral reef die-offs and other impacts of ocean acidification and ocean heating all the way up the food chain. In other words, even if these SRM approaches successfully halted global warming, we'd still be left with dead oceans.

The most talked about SRM concept is aerosol injection into the stratosphere. The authors of the wildly popular Freakonomics books, Stephen Dubner and Steven Levitt, featured the aerosol injection concept on Dubner's radio show.

As much as I enjoy the books, Dubner is neither a scientist nor an engineer (and doesn't pretend to be). On his radio show, he featured an SO_2 aerosol dispersal concept consisting of hoses and pumps held aloft by a chain of tethered, helium-filled balloons. The concept was referred to as a "garden hose" reaching 100,000 feet into the stratosphere. Once deployed, SO_2 would be pumped from the earth's surface and dispersed in the stratosphere

One of the goals of the Freakonomics franchise is to encourage people to think differently, (like a freak?), and in the case of SRM concepts, to appeal to people to not immediately dismiss the concept out of hand. I understand "thinking differently," but I'm afraid that reference to a 100,000 foot "garden hose" reaching into the stratosphere, while entertaining, trivializes a complex technical problem to the point where some people might think, "this is crazy," while others might think "this is easy." (After all, they heard a smart guy talking about the idea on the radio, right?)

Gas-filled balloons are capable of lifting significant payloads to incredible heights. Daredevil skydiver Felix Baumgartner demonstrated this when a helium-filled balloon hoisted his 2,900-pound capsule to 128,000 feet, the altitude from which he performed a successful free fall back to Earth in 2012. Baumgartner's balloon, however, wasn't tethered, and no tethered balloon or chain of balloons has ever flown to this altitude, ever. Nor has anyone demonstrated the ability of a chain of tethered balloons to remain aloft, intact, for any period of time in a pretty hostile environment.

There are other, more proven techniques (e.g., high-flying, modified, commercial aircraft) that could be used to disperse SO_2 in the lower stratosphere, but those techniques were barely mentioned, perhaps because they were less entertaining than the image of a "garden hose" rising like Jack's beanstalk, nineteen miles into the sky. Highlighting the balloon/hose concept, I felt, was a like giving equal time to a flat-earth proponent on a documentary about the cosmos.

Some authors have identified a number of potential problems, i.e., unintended (and unfavorable) consequences, associated with dispersing sulfur compounds in the stratosphere. Some question the longevity and effectiveness of the aerosol particles and worry about their tendency to coalesce and fall back to Earth as acid rain. Others are concerned that these aerosols could possibly disrupt mass and heat transfers in the stratosphere, interact with ozone, increase ocean acidification, and impact vegetation as the acid particles settle out over time. There could also be differential regional impacts. Some objections are also philosophical: "Does it make sense to pollute our atmosphere even more to try to counteract the damage we've already done by polluting too much."

The point is, we don't know a lot about all the impacts that aerosol dispersals might have on the stratosphere. A lot of actual field work (small- and large-scale experiments, in addition to laboratory work) would have to be done before even suggesting large scale SRM testing should be conducted. We just don't know enough. The one advantage SRM actions might have if they're ever proven to be effective, safe, and economical, is that they could be employed to a greater or lesser extent depending upon how much cooling is required. The longer we continue burning fossil fuels, the more intrusive the countermeasures will have to be.

In addition to **not** addressing ocean acidification resulting from the buildup of CO_2 in the atmosphere, as noted above, SRM technologies would also be encumbered by all of the same governance questions as those related to a carbon tax or carbon capture.

1.   Who's going to pay for it?

2.   How much data is necessary before large (or even small) experiments are considered prudent or even permitted?

3.   Who or what organization will decide what can and can't be done; what should and shouldn't be done?

4.   If there are negative consequences in some regions, how will the "injured" parties be compensated?

5.   If negative consequences do materialize, what will be the effects of abruptly stopping the aerosol interventions?

What Will It All Cost

Some groups have estimated the cost of addressing climate change to be in the range of a trillion dollars a year for a long time (for all countermeasures).   They point out that even this massive expenditure pales in comparison to the ultimate cost of not doing anything.   Some studies suggest a 3-to-1 return on a full-scale investment to mitigate climate change. One of the problems with this kind of analysis, however, is that the majority of the costs for CO_2 countermeasures are incurred in the present while the majority of the benefits (large as they may be) are in the distant future. And, in all honesty, I don't know what's included in the trillion dollar per year price tag or the calculated savings. However, it may be useful to put this level of expenditure in perspective.

First and foremost, a trillion dollars is a **lot of money**. When members of Congress want to highlight a major expenditure they oppose, they often resort to describing the expenditure in terms of the height of a stack of one-dollar bills. In this case, a trillion one-dollar bills would reach more than a quarter of the way to the moon. Does that help? I didn't think so.

Closer to home, the first of the COVID-19 economic rescue packages passed by Congress was $2.2 trillion, the second was $3.5 trillion. Still not clear?

Maybe this will help. The U.S. funds about one-quarter of the of United Nations' programs – we contribute roughly $10 billion/year. Ten billion dollars is just 1% of one trillion dollars. Each time the UN appropriation comes up for a vote, opponents complain that UN programs are not **exclusively** advancing U.S. interests. Can you envision a U.S. Congress approving an annual expenditure one hundred times our annual UN funding for programs whose benefits extend to the whole globe?

Technology in Hand vs Research in the Lab

I believe in technology. I believe in research. I fully expect the fruits of those efforts to yield improvements of all kinds: efficiency gains, cost reductions, better materials and processes, and so on. But we're in a battle right now to dramatically curtail the worldwide production of CO_2 and phase out the consumption of fossil fuels for all purposes. Innovative technologies, such as the artificial leaf and others, may be coming, but it's unrealistic to expect that an embryonic technology or something we don't know anything about in 2022 is going to miraculously transform our entire world's energy economy within the next 20 or 30 years. In addition, these new technologies seldom turn out to be quite as cheap, reliable and effective as they may appear in the laboratory. The real world has a way of taking the luster off most bright, shiny objects.

> *Innovative technologies seldom turn out to be quite as cheap, reliable, and effective as they appear in the laboratory. The real world has a way of taking the luster off most bright shiny objects.*

[*Input] I'm certain there are numerous developments in the lab or technologies on the drawing board, that I've failed to mention or adequately highlight. My apologies. If anyone knows of a technology that could alter the trajectory of the world's energy economy in the next 30 years, please send me a description, projected costs, status of development, etc. through the website, BlueOasisNoMore.com, for consideration to be included in the second edition of the book. (Also, please explain why the technology you're highlighting has been kept a secret up until now). [*Input – End]

In addition to the time required in the research and development phases, it also takes time for industry to ramp up the production of new technologies. Those new products and those scaling activities often expose other deployment hurdles. The biggest hurdle of all, however, may be the penetration of a new, more expensive technology into a market that favors the status quo. Just as most consumers will be unwilling to abandon their major (fossil-fuel-consuming) appliances and other durable goods, all new technologies will be attempting to enter a market where existing industries have a dominant position. Those players aren't going to relinquish their dominance without a fight. We may think markets are smooth and continuous, but in fact, they're viscous, sticky, and lumpy . . . even when a better mousetrap comes along. ["Viscous, sticky and lumpy" isn't just a description of markets; it's a description that explains a lot about most things in the real world. But that's a different book.]

Action Items

Research will continue to be performed on every aspect of global warming and possible countermeasures. Many will yield promising results. Because research dollars are limited, preserving planet Earth should be given a high priority.

1. Terminate federal funding for fusion-power research.

2. Defund **manned** Mars and moon mission activities.

3. Reprioritize all budget allocations (domestic spending, foreign aid and defense spending) to maximize funding for planet survivability investments. These actions would be more symbolic than substantive but might help the public understand the urgency of our CGI.

---

17 THE ROLE OF THE MEDIA:
TIME MAGAZINE'S 2019 SPECIAL CLIMATE ISSUE

"If wishes were horses, beggars would ride."

Scottish Proverb, 1628

This chapter is a review of TIME magazine's Special Climate Issue (9/23/2019). This was not a fun chapter to write.

I feel closely aligned with those in this activist-space who are trying to move the U.S. (and the rest of the world) to a carbon-free future. As a result, I feel more than a bit uncomfortable criticizing the editors and many of those who contributed to this special issue. On the other hand, I feel that activists encouraging actions to combat global warming need to be held to at least the same level of scrutiny as the global warming naysayers.

In this chapter, I'm specifically discussing the content in the 2019 special climate issue of TIME, **but my broader concerns extend well beyond this one magazine to what I see as overly optimistic reporting,** bordering on delusional**, in most media.** I believe that much of what's in the news shows we watch, the papers we read, the radio programs and podcasts we listen to, and the magazines and books we read, all serve to raise our concern, but at the same time, lull us into a dangerous complacency. The media is not wholly responsible for how the broader society reacts to its reporting, but the general public has become numb to the constant drip-drip of global warming stories. I believe this is because these stories most often consist of two parts: 1) things are dire, but 2) if we do x, y, z, we can "beat" this. This issue of TIME is a perfect example of this dual messaging and is one of the reasons why I gave this book the stark and perhaps shocking title I gave it.

TIME's Special Climate Issue has the cover story essay: "2050 – How Earth Survived." Since my proposition is that mankind is on a superhighway to "hell *and* high water" in the second half of this century, I'm devoting an entire chapter to examining how, with the same information, at least some of the TIME contributors came to a rather different conclusion. Few readers will even remember this "special issue," but it made quite an impression on me. This issue of TIME convinced me that I needed to write the book, this book, that had been rattling around in my brain for over a decade.

First, I should make clear, I agree with a lot of what's in TIME's special climate issue with the exception of four fundamental perspectives:

1. Their unsubstantiated (and unwarranted) optimism about the future,
2. Their inflated expectations that technology will save us,
3. Their unwavering confidence that "youth activism" will turn the tide, and,
4. Their failure to offer a comprehensive list of specific actions that support the premise they're trying to advance.

In addition to these four major complaints, I had a bunch of minor issues that gave me pause. In theory, I should focus on just my four major objections and "not sweat the small stuff." In this case, however, I don't feel like I have a choice: ANY article, ANY book, ANY speech, ANY conversation, ANY interview, ANY anything - that doesn't communicate the magnitude, the urgency and the catastrophic nature of the problem, may be contributing to complacency and inaction.

People want to believe that the medicine to cure this illness won't really taste that

> **ANY article, ANY book, ANY speech, ANY conversation, ANY interview, ANY anything - that doesn't communicate the magnitude, the urgency and the catastrophic nature of the problem, may contribute to complacency and inaction. People want to believe that the "medicine" to cure what ails us won't really "taste" bad. (They're wrong. It will.)**

bad. (They're wrong, it will.) If there's even the slightest suggestion that some technological breakthrough is going to conquer this demon, people may be less inclined to act immediately and less willing to make the changes, support the policies and spend the resources to do what needs to be done, right NOW. Consequently, some of my comments in regard to the TIME articles may appear to be petty, but I'm raising them nevertheless in an effort to help people NOT become complacent, but also to be vigilant, cautious consumers of the material they're reading (just as I have urged you to be in reading this book). So here goes . . .

The title article for TIME's special climate issue is an essay by well-known environmental writer, Bill McKibben. Mr. McKibben's piece is an imaginary retrospective look back from the year 2050 at what we "did" (past tense) to slow the onset of global warming impacts. It's intended as a guideline for the things we need to do in the real world beginning yesterday to address global warming. If you read the TIME issue from cover-to-cover, you can't help but come away with a somewhat hopeful feeling. That's certainly what the editors intended.

However, the title article by Mr. McKibben makes **exceedingly clear** that the 2050 earth he envisions is changed forever. Even after implementing all of the policies and actions he fantasizes taking place between 2020 and 2050, he concludes that this is a time in earth's history that's **"wretched, . . . which is [at least] considerably better than catastrophic."** (His words, not mine.) Unfortunately, the title the essay, "How We Survived Climate Change", doesn't exactly prepare the reader for the "wretched" century Mr. McKibben is rightly warning us about.

At the heart of this logical disconnect, my first objection relates to the use of the word "survived" in the banner title for the issue. When most people hear or read that word "survived," I think the image that flashes through their mind is a survivor who's up, walking around and doing all the things they did before their trauma. Technically, of course, that's **NOT** what "surviving" means. A soldier can lose both

> *People remember the headlines,*
> # "Earth Survived,"
> *not the fine print,*
> "It was wretched."

legs, one arm and the sight in one eye and still be designated a "survivor."

People remember the headlines (Earth Survived), not the fine print (It Was Wretched). It's almost as if the TIME editors inadvertently buried "the lead." Instead of "How Earth Survived,"

They could have run with "Earth 2050: Wretched – At Least Not Catastrophic."

Mr. McKibben is optimistic in terms of what he thinks will take place in the next 30 years, but if you read his article closely, he's not overly optimistic that mankind will actually avoid some exceedingly tough times ahead. Many of the subsequent articles in this issue of TIME were decidedly more upbeat than the one by Mr. McKibben. Reading them made me wonder how Mr. McKibben reacted when he read some of them that suggested that everything was going to be just fine . . . not exactly "wretched." By taking an upbeat, "if-we-get-our-act-together" approach, I'm afraid the TIME issue is sugar-coating a description of just how hard and how expensive "getting-our-act-together" is going to be and how much resistance those realities will foster.

I'm arguing that there are real obstacles that are going to prevent us from getting on the path to really confront global warming. Many of the contributors to this special issue acknowledge these difficulties, but gloss over them in favor of optimism. In this book I'm trying to shine a bright light on some of those pesky details; to do otherwise, in my mind, would be naïve and reckless.

Climate action advocates never intended for their upbeat assessments to contribute to public complacency surrounding this issue, but I believe they have. What other conclusion could you possibly come to after reading that researchers are working on advanced batteries, advanced generating technologies, advanced carbon removal technologies, . . . advanced everything! "Advanced technologies are going to fix the problem, right? Things aren't going to be so bad!" Even if that wasn't what most of the contributors were actually saying, that's what most of the general public probably took away from the totality of this edition.

In "The Sands of TIME," the magazine's Editor-In-Chief congratulates his publication for being a leader in raising the public alarm back in their first climate change issue 30 years ago. This is probably a well-deserved, if self-ascribed, accolade. Awareness has been raised and renewable energy has indeed made great strides since 1989, BUT, 30 years on, the critical question is not: how much has been accomplished? The critical question is: how much has been accomplished in combating global warming COMPARED TO how much needed to have been accomplished by this time?

I'm afraid the answer to that question is: painfully little. Yes, there have been significant technical and economic advances as I've previously described. Progress has been made in improving the efficiency of photovoltaic systems and especially in reducing the cost of the solar panels themselves. There have also been improvements in the efficiency of wind turbines and reductions in the installed cost/kW of wind generation.

> *A lot of progress has been made in the past 30 years. But the critical question is: how much has been accomplished in combating global warming* **COMPARED TO** <u>how much needed to have been accomplished</u> *by now?*

LED lighting has emerged as an enormous (though largely unheralded) advance. Lithium-ion battery technology has made monumental leaps forward which has contributed to making electric vehicles almost commonplace, at least here in Southern California. But none of these technical advances or successes in the marketplace was fundamentally driven by a willingness on the part of the general public to make sacrifices on behalf of the environment. As I argued in Chapter 13, these were mostly economic choices where the benefits were greater than the cost **with only a relatively minor consideration of CO_2. We will only finally know that concerns about global warming have really taken hold when the general public endorses climate policies that take money out of their own pockets for hard-to-quantify, societal (i.e., not personal) benefits, most of which will be realized by generations yet unborn.** That hasn't happened.

Inflated Expectations for Technology Fixes

The TIME issue purports to "explore the essential role of innovation in solving the crisis." As an engineer, I generally believe that innovation is good, but I fear that too many people are expecting a benign, technological "silver bullet." As I argued in Chapter 16,

1. Research and development take time,

2. Deployment and market penetration of new technologies take time, and,

3. We don't have time.

Not only don't we have time, we can't wait. We need to use the tools we have in hand today and act now. We need to treat global warming with the urgency of the Manhattan Project. (Chapter 4)

I commend the magazine for not feeling compelled to present the views of climate deniers. Clearly, the onus is on the naysayers to present credible arguments for their positions. Until they do or at least try, they haven't earned a place at the table. They've sidelined themselves and don't have the right to have their views aired by reputable sources like TIME.

The Editor-In-Chief asserts that "[It] (i.e., doing what's necessary to combat climate change) is a moment we can rise to . . . [by taking] collective action." Again, my complaint is that this is aspirational optimism unsupported by real world observation. It sounds nice, but before the discerning reader can accept this assertion, the editor (and/or the contributors) have to provide credible evidence for their optimism. I've pointed out how and why we have not risen to the occasion yet and that we're exceedingly unlikely to do what actually needs to be done soon enough to make a difference.

Mr. McKibben, writing from a 2050 perspective, observes that the "most intense phase [of the climate fight] may be in our rearview mirror." This 2050 perspective, he asserts, is because Americans understood that they were at risk, that technological advances have yielded cost effective solutions, that Green Parties (or principles) in Europe, the U.S. and around the world made significant gains in the 2020s, that tax subsidies for the oil and gas industries have been redirected to green energy options, that American homes have replaced oil and gas burning systems and appliances with electric options, and NIMBY has given way to "vibrant urbanism."

These are all good things and I would agree all are likely to happen at least to some degree. However, modest improvements in this case won't cure the patient. In this book I've explained why, by 2050, the worst consequences of global warming will have just begun to take hold and that the most intense phases of the climate fight will most certainly not be in our rearview mirror. None of the measures Mr. McKibben describes will stop global warming, nor does he claim that they will. In fact, he states that the earth's temperature is continuing to rise in the 2050 world he imagines. This is why he concludes that life in 2050 will be "wretched." However, he also believes we will have turned a corner by then. I've explained why that cannot possibly be so. I too would like to be optimistic, but the rate at which the earth is warming will increase every decade for the foreseeable future. The best we can hope for is that the rate of adding CO_2 to the atmosphere will slow, but the global average temperature and the rate of warming will both continue to increase.

Kicking the Can Down the Road

Mr. McKibben's article precedes an article by former Vice President Al Gore. In keeping with his character, Mr. Gore is upbeat, but acknowledges that because of the CO_2 load the atmosphere is currently carrying, **"more damage is now inevitable *no matter what we do.*"** (Emphasis added). Please reread that last sentence. He also acknowledges that, "False hope is a form of denial," but then, nevertheless, continues, "I remain optimistic . . ."

What's the basis for his optimism? It's the same as that expressed by Mr. McKibben in his essay and Dr. Goodall in her article. Their optimism is based on youth activism.

They're optimistic because of the engagement of young activists. I too am heartened by their involvement, of course, but I think **relying on youth activism to solve the global warming crisis is flat out wrong.** It's wrong on four counts:

1. First, putting such an enormous emphasis on the engagement of young people feels a little like an abdication of responsibility by those of us over 30, a surrender. It's as if we're saying, "Since we created the problem and couldn't (or wouldn't) fix it, it's up to you, young people. You fix it; we're counting on you." Openly counting on young people to fix what we haven't been able to, feels a little like letting ourselves off the hook. Because the buildup of greenhouse gases in the atmosphere is effectively cumulative, making a credible dent in CO_2 emissions is going to take a concerted effort by every generation for at least the next 100 years -- not just today's youth.

2. Second, it will take time for young people to establish themselves in all the levels of business and government where they can influence decision making in favor of sustainable policies, but we don't have the luxury of time. To have any hope of altering the trajectory of climate change, drastic changes have to be made right now. Yes, young people can be part of the pressure campaign to redirect public policy discussions, but there's no substitute for

being "in the room where it happens" and right now, young people are not in those rooms.

3. The third way in which I feel it's wrong to rely too heavily on youth activism to solve the global warming crisis relates to idealism. I know this may sound hard to believe, but young people eventually get older. As they do, they typically assume greater and greater responsibilities in their personal lives. Often, that transition strips away idealism. It doesn't have to, necessarily, but more often than not, it does. As young singles become young families, mundane obligations (like paying the mortgage or rent, putting food on the table and kilowatt-hours into the EV) change people. Is there something about today's young folks that will make them any more receptive than earlier generations to pay more for everything they buy now in order leave the world a better place for someone they don't know, decades in the future? I haven't seen it.

4. And finally, young people are painfully ineffective in the political arena because they lack the two most important ingredients that could make them effective: money and votes.

It may be worth elaborating on this fourth reason for not relying too heavily on youth activism to defeat global warming. The vast majority of young people don't have a lot of money (i.e., political lubrication) and even more importantly, they DON'T VOTE! Let me put that a bit more diplomatically: **historically, voter participation by young people has been painfully low.** Even the most engaged young person will have to acknowledge this disturbing, historical fact.

But wait. All indications are that young people DID, in fact, show up in greater numbers in the 2020 election. This was a **monumental** change and I'm heartened to have seen it. However, the 2020 election was NOT exactly an endorsement that the young "have got this." The shift away from a president who was a climate change denier back to a president who embraces science could hardly be characterized as earth-shaking. The sad fact is that even if climate change was not the primary issue for most voters, just a couple of percentage points shy of 50% of the electorate was willing to support a candidate who was openly hostile to environmental action.

Yes, young people can bring enthusiasm and passion to the fight against global warming, but their active participation, while necessary, **is not a sufficient condition.** They have to get their parents and peers to sign on as well. Combating climate change is not one of those issues where random or occasional victories will get the job done. It will require active engagement by both the young and the once-young over their entire lives, regardless of how painful it becomes.

My lack of confidence that young people will be able to do enough to save our planet may not be what committed young people want to hear, but all I can say is: **"prove me wrong."** Paraphrasing Cuba Gooding Jr. and Tom Cruise in the movie, <u>Jerry Maguire</u>, "Show Me the Money" [aka, results].

Mr. McKibben suggests that non-carbon producing power sources other than solar and wind may be on the horizon in the transition from 2020 to 2050. He notes: ". . . researchers continued to work to see if fusion power, thorium reactors or some other advanced design could work."

It's true that research is being done today on fusion power, alternative nuclear fission reactor designs and advanced nuclear fuel cycles, but I cannot, for the life of me, figure out why or how this offhand comment adds anything of substance to his argument that "earth survived." He makes no claims that any of these technologies have displaced a single kilowatt-hour of energy extracted from coal or natural gas in the thirty-year period from to 2020 to 2050. As I argued in Chapter 7, there are a variety of reasons why advanced nuclear reactors will not be deployed quickly enough to curtail global warming. In Chapter 16 I gave an even longer list of technical reasons why fusion reactors are unlikely to ever produce a single kilowatt-hour of usable energy before or after 2050. Mentioning fusion power and thorium reactors in a discussion about turning the corner on saving the planet by 2050 may suggest an openness to new technologies, but I fear that the average reader of TIME magazine, especially given the esteem in which Mr. McKibben is held, might conclude that these new power resources are on the brink of being deployed and that all will be well.

Mr. McKibben also asserts that there's been a "marked shift in public demand for action." I certainly agree and understand that public polling reflects more support in favor of actions to combat global warming, but I have serious doubts that most of those responding in the affirmative really know what that might entail. And furthermore, once they fully understand what they will need to do, I'm afraid that public support for specific actions will fade or even morph into outright opposition. As I mentioned earlier, look no further than how the public has responded in the past to the suggestion of increasing gasoline taxes to discourage consumption. And now, in the aftermath of the Covid-19 pandemic, look at the public (and political) reaction to five- and six-dollar gasoline! Where in this picture is there a shred of evidence supporting the hopeful, optimistic conclusion that we will "rise to the occasion." I don't see it; in fact, I see the opposite.

Ladies Night

The editors at TIME have chosen to highlight the contributions of fifteen women who will "save the world." They also report that "the UN estimates that 80% of those displaced by climate change are women." They go on: "Given their position on the front line of the climate change battle, women are uniquely situated to be agents of change . . ."

This seems to me to be a case of enormously fuzzy thinking and/or blatant pandering to their female readers.

While I generally believe that the world would be better off if all elected governmental leaders were women (**now look who could be accused of pandering** – but no, trust me, I've said this for years), combating global warming is not a "gender thing." Injecting gender into the discussion strikes me as an unnecessary distraction. In addition, I highly doubt that any of the fifteen women profiled in the magazine were ever displaced due to climate change. And even if they were, being displaced or being women doesn't make them effective agents of change. **Hopefully being smart, educated, observant, well-informed, honest, logical, and reflective were the attributes for which these remarkable individuals were selected to discuss climate change policy, not their gender.**

I turn now to the article by Jane Goodall, whom, like <u>Silent Spring</u> author Rachel Carson, I greatly admire. (Who doesn't?) Quoting Dr. Goodall, "in order to <u>slow down</u> climate change, we must solve four seemingly unsolvable problems:

1. We must eliminate poverty,
2. We must change [our] unsustainable lifestyles,
3. We must abolish corruption, and
4. We must think about our growing human population."

First things first. Remarkably, (and to her credit) Dr. Goodall is the only contributor in the climate change issue who explicitly raised the issue of "our growing human population." Everything we're doing or will do to reduce carbon emissions will be even harder as our world population grows. Regrettably "circumstances" (i.e., famine, war, and disease) may address this issue in a painful way long before we homo sapiens (the "wise ones") do.

Dr. Goodall identifies four "seemingly unsolvable problems" to just slow climate change. One of those actions is to make significant changes to our unsustainable lifestyles. Because she was allotted only one page in TIME's climate issue, she's to be forgiven for not explaining what those changes would have to be and how they would be made. However, as I've argued, people don't often respond well when they're asked to change and they especially don't like being forced to change.

One of the other requirements Dr. Goodall identifies is eliminating poverty. Setting a goal of raising real human beings out of abject poverty is absolutely the right thing to do. Perhaps you know the expression: "There but for the grace of God go I." The TIME issue is filled with photos of people living in conditions that are extraordinarily hard – conditions that would be barely tolerable from a Western perspective. But here's the tricky part: raising the standard of living for literally billions of people, besides being "seemingly unsolvable," would also be catastrophic for the atmosphere.

Raising people from poverty means improving diets, adding infrastructure to provide new services (schools, shops, roads, housing, sewers, clean water, electricity, medical care, etc.), providing more consumer goods, more of . . . everything. After all, that's what poverty is . . . i.e., the state of lacking pretty much everything. Raising the standard of living for billions of people around the world to that enjoyed by most Europeans (much less Americans) would dump billions upon billions of tons of additional CO_2 into the atmosphere every year. (This is essentially what's been happening in China as that country has moved slowly up the "wealth" ladder . . . and the "per capita CO_2 emissions ladder.")

Mind you, I'm not saying that we (and especially the rich Western nations) shouldn't work to raise the standard of living of the entire world's people. We should. They are, after all, our fellow human beings. I'm just pointing out, as have many others, that doing so will increase the difficulty of creating a future free of human-activity-driven CO_2 emissions.

Dr. Goodall's four requirements will, in her words, just "slow" climate change. They won't enable us to "beat" it. Furthermore, and regrettably, I believe her characterization of those four requirements as "seemingly unsolvable" may be an understatement.

If by now you've concluded that I'm unhappy with a number of the articles in the TIME issue, you won't be surprised that I was also upset by the article focusing on Europe. The central theme in the offering by journalist Ciara Nugent was about removing CO_2 from the atmosphere. A more powerful climate discussion about Europe might well have focused on the fact that per capita CO_2 emissions in the EU are HALF of what they are in the U.S. with no degradation in the quality of life. That could have been truly enlightening. Instead, her article begins by featuring a project on Wallasea Island, east of London, to restore a wetland, "[a process that] holds a key to our future." Good news, right? But it also reports that over 90% of the wetlands in this area have been lost in the last 400 years and, worse, 35% of the entire world's wetlands were destroyed between just 1975 and 2015! Not good news.

It seems to me that if wetlands really are a "key to our future," the primary focus of the article should have been on how to stop the ongoing destruction of wetlands BEFORE it happens and the progress (if any) that's been made toward that end. Isn't that the first principle of triage, i.e., stop the bleeding?

The article also highlights technologies to extract CO_2 from the atmosphere. One of the start-up companies highlighted is Climeworks which extracts CO_2 from the atmosphere and then sells it. The company is also experimenting with the storage of CO_2 in underground rocks. Most of the CO_2 captured by Climeworks, however, is sold commercially for use in greenhouses, soda manufacturing and biofuels production – all processes that ultimately release their CO_2 back into the atmosphere. Those activities yield no net reduction in atmospheric CO_2. My complaint, however, is not with Climeworks. My complaint is with TIME for failing to point out this critical fact. For me, this editorial choice raises a huge red flag. It makes me wonder if they've "edited" all of the articles to support the pre-ordained premise that earth survived.

The next article, "Paper Straws Alone Won't Save the Planet," by Michael E. Mann, argues that "appearing to force Americans to give up . . . things central to the lifestyle they've chosen to live is politically dangerous." I agree. He goes on: "The good news is we have tactics to bring environmentally friendly (and non-lifestyle-disrupting) options to fruition: pricing carbon emissions and creating incentives for renewable [carbon-free] energy and reduced consumption." Dr. Mann is absolutely right about the value of carbon pricing as a tool to cut emissions and reduce consumption, and of incentives to promote desired behaviors. However, he's absolutely wrong to characterize meaningful and effective carbon pricing as "non-lifestyle-disrupting."

"Pricing carbon emissions" . . . Dr. Mann is talking about a carbon tax (Chapter 13, Economics). He apparently wants to avoid calling it a "tax," but that's what it is. My more fundamental complaint, however, is Dr. Mann's suggestion that a carbon tax will be "non-lifestyle-disrupting." He's right that a carbon tax will be non-lifestyle-disrupting if it's too low. But if it doesn't disrupt lifestyles, i.e., consumption choices, it won't reduce CO_2 emissions. If we imposed a $500 environmental penalty on the sale of every new Hummer and distributed those revenues to all Prius buyers (rewards), the impact on the sales of each vehicle type would be imperceptible; there'd be no change in emissions. But if that environmental penalty were $50,000, new Hummer sales would plummet. Emissions from new Hummers would cease. In the area of carbon pricing, "size matters." That's the whole point of a tax: to discourage consumption. It has to be large enough to change consumption choices -- which is the definition of lifestyle disrupting.

A number of the contributors to the TIME issue made the case, as do I, that responding to the threat of global warming will require a wholesale transformation of the world economy from fossil fuels to renewable energy sources. If Dr. Mann doesn't think that will be disruptive, I don't know what would be.

The author also implores us to "never underestimate the resourcefulness of Americans when there's a dime to be made." Again, he's absolutely right, but unfortunately, that resourcefulness is a double-edged sword. We must also never underestimate the depths to which some among us will sink to make that same dime: doctors who scam Medicare, disreputable contractors, Purdue Pharma or [*insert here the name of your "favorite" most-scurrilous, corporate villain*]. In other words, never underestimate the resourcefulness of Americans to exploit loopholes in regulations to make a dime or figure out a way to do what they want to do regardless of the consequences.

In late December, our local paper carried the headline, "Solar Power Spurs Housing Permits." I was intrigued and expected to read an article about how new home buyers were putting a premium on solar-equipped homes.

Silly me. The article explained that new regulations requiring solar panels on all new homes permitted after the first of the year were about to take effect and that home builders were filing for building permits in order to avoid having to add solar panels on their new homes. Solar panels would drive up the sales price and potentially make their new homes harder to sell even though the new owners of solar-equipped homes would enjoy lower utility bills for many years into the future.

I also remember a circumstance that developed many years ago, after an energy shock in the 1970s. A new federal program was developed to incentivize homeowners to add insulation to their homes by providing an insulation rebate. The guidelines for the new program weren't written very carefully. They didn't restrict the rebate program to existing homes, so builders of new homes began reducing their costs by leaving insulation out of their new construction so that the new buyers would qualify for the Federal rebate. Unfortunately, of course, retrofitting insulation after a house is already built is a lot more expensive and much less effective than installing insulation during the construction process.

The Farmers Will Save Us

In Justin Worland's article entitled, "The Climate Caucuses," a reference to the presidential primaries at the time, the author quotes an Iowa farmer who shares his opinion: "Farmers and rural Americans, that's who's going to solve [climate change]." With all due respect to a fellow farmer (i.e., I have 0.003 acres of tomatoes in my backyard), farmers are NOT going to save us.

Farmers most assuredly have a role to play, but they are not going to solve global warming. Significant numbers of wind turbines (and some solar) will be installed on their rural lands and modern farming techniques will undoubtedly preserve precious farmland and sequester some carbon, but the vast majority of the grain, vegetables, fruit and livestock that farmers produce remove CO_2 from the atmosphere only temporarily until we give it back when we consume the farmers' bounty.

Farmers are important, but farmers are not going to build electrochemical battery energy storage facilities, not going to build EVs, not going to electrify railroads, not going to fix natural gas pipeline leaks, not going to build advanced nuclear plants, not going to build pumped hydroelectric storage plants and all the rest.

Additionally, like the rest of us, farmers have only one vote each to elect politicians who can end subsidies to oil and gas producers, enact a carbon tax, negotiate with governments around the world, ban plastic bags, and adopt other planet-friendly policies. Because every aspect of human existence has implications for CO_2 emissions, farmers are most definitely players in the global warming story, but we all are.

Protecting the Most Vulnerable

The offering by Adrienne Hollis reminds us that "climate change is a public health issue." She makes clear that the path forward needs to include policies and programs that protect and buffer the impact on our most vulnerable communities. Dr. Hollis is referring to programs to help the most vulnerable adapt to global warming and deal with governmental policies that will affect their financial stability (e.g., a flat carbon tax). She's right, but not everybody in the country is onboard with government programs that help others at their expense. Protecting have-nots from the effects of global warming will cost money. The opponents will see such programs as a redistribution of wealth, i.e., **their** wealth. Saving the planet in a compassionate and just way is going to rekindle old political arguments and those arguments will detract from public support for such programs. In other words, the alignment necessary to make real progress (Chapter 14) will be disrupted. Nobody said this was going to be easy. Well, actually, that's not true.

When "B" Does Not Follow "A"

The article by Andrew Blum suggests that we can "innovate our way out of this mess." He discusses advances in renewables, grid interconnections, energy storage, next-generation nuclear plants, and carbon sequestration, **all of which is true.** But this is an example of making broad, non-specific statements which are then used to generalize to a bigger conclusion that sounds logical. In this case, Mr. Blum concludes: "Life could not only go on – it could go on more or less as it has."

I don't agree . . . unless what Mr. Blum is saying is that life will go on more or less as it has even if people are paying three or four times more than they do now for the energy component of everything in their lives (i.e., not just gasoline and electricity, but also the energy content in their food, water, consumer products, packaging, moving goods around, vacation travel, everything). For all but the rich (and perhaps even for some of them) when people get squeezed financially, they have to make choices. Those choices in 2075 will be quite different from the choices their parents or grandparents made in 2020.

As I've argued in many of the prior chapters, life as we've known it has been thriving off the stored resources of freshwater, rich topsoil, energy-dense fossil fuels, vast forests, immense biodiversity, abundant marine life, etc. We've been partaking of the "low hanging fruit" for the last 100 years. The reason it's called "low hanging fruit" is because it's the easiest, cheapest, and most abundant "fruit." Everything else is a lot more expensive and harder to obtain.

The energy economy we've built over 150 years never included the effects of CO_2. Rebuilding that economy with carbon-free resources will be a monumental

> *The energy economy we've built over 150 years never included the effects of CO_2. Rebuilding that economy with carbon-free resources in three decades will be a monumental and expensive undertaking.*

undertaking. If we do what's necessary to respond to global warming in a time frame that allows humanity to avoid its worst consequences, it's going to be costly. How could anyone conclude that transitioning an entire world economy with nearly eight billion people away from fossil fuels to carbon-free energy sources could be accomplished in 30, 40 or even 50 years without life-altering changes? Yes, we will still live in houses or apartments and, yes, we will still eat food, and yes, we will move around in vehicles of one kind or another, but what we do with our time and our money, how we are affected by nature and how we interact with our fellow citizens will be dramatically different. Life will NOT go on more or less as it has.

My concerns about Mr. Blum's article, however, are not limited to his unbridled belief in innovation as humanity's savior. In his article he states that "energy storage will increase by a factor of 10 by 2023." Including this fact about energy storage is an explicit recognition by Mr. Blum that energy storage is a critical component of transitioning to a fully green economy. I discussed this in Chapter 9. However, by highlighting a 10-fold increase in energy storage by 2023 (which is no doubt true), I believe he's leading the reader to conclude that great strides are being made in the deployment of the bulk energy storage capacity needed to phase out fossil fuels. That's not true.

This is an example of reporting a **fact** in a way that may lead the casual reader to an erroneous conclusion. Here's what I mean. What if, for example, Tesla reported that it sold 10 EVs in its first month of sales and 100 in its second month? It would be factually correct to report that sales had increased 10-fold in just one month. Reporting that increase might lead a reader to conclude that Teslas were taking over the car market when in fact, the sale of 100 vehicles might represent just 0.007% of all monthly vehicle sales in the U.S. Without qualifying remarks, reporting substantial percentage increases from a base number that's small is almost certainly intentional. Intentional or not, authors should think about how their readers may interpret what they've written.

Most of the storage capacity that's currently being deployed will provide stability to a grid with large quantities of intermittent energy generation. It's not being deployed to accomplish a wholesale displacement of fossil-fuel generating stations and the CO_2 they produce. And, as I pointed out in Chapter 9, recharging energy storage with either renewable or fossil fuel energy sources at a time when any fossil fuel energy is flowing into the grid will actually increase CO_2 emissions compared to what they would have been without the storage.

If our goal is to decrease the use of fossil fuels, what matters is not the percentage increase in storage capacity from year-to-year, but rather the year-to-year increase in the percentage of all electric energy that customers consume that's delivered from energy storage installations. The latter gives you something that might actually be important in regard to displacing fossil-fuel consumption (provided, of course, that the storage capacity was recharged using carbon-free generated energy and fossil fuel contributions to the grid at the time were minimal).

I'm not offering these opinions about the article by Mr. Blum and the TIME editors to be contrarian. Instead, I'm pointing to these examples in hopes that readers may be more discriminating in how they consume information, even from trusted sources like TIME and its contributors. I'm also hoping that editors will exercise increased oversight over what they publish and that writers always how their readers will interpret what they've written.

So, What Do We Need To Do?

My last major criticism of the TIME issue is that it doesn't lay out the concrete actions that "were" taken to ensure that Earth survived.

Most of the contributors to this special climate issue describe areas in which progress could be made to combat global warming, but they're all pretty general. More than 100 pages into the issue, under the heading, "Road Map," TIME finally provides what it calls "A 30-Year To-Do List." The only problem: it's not really a To-Do list.

My idea of a To-Do list is a list of specific tasks that have to get done in some time frame. For example, a Saturday To-Do list might read: 1) pick up laundry at the dry cleaners, 2) wash the car, 3) buy a birthday card for Uncle Willie, 4) work out, etc. These kinds of lists help us set priorities, plan our day, and make sure things don't fall through the cracks.

TIME's "To-Do" list doesn't list any specific actions. Instead, it's a list of areas that need attention: increase energy efficiency, support trees, change agriculture, remove carbon from the atmosphere, grow renewables, phase out fossil fuels, "chart a path on nuclear." I'm sorry, but **this could only be characterized as a "To-Do List" in the same way "Get More Points Than the Other Team" could be thought of as a "game plan."** Neither is actionable; neither increases your chances of "success."

Even the inset at the bottom of the page (i.e., "How We Get There,") isn't specific. It refers to government, corporate and individual commitments, and the need for innovation, but doesn't begin to tell the reader what those commitments need to be, what areas are in need of innovation or any inkling of "how to get there."

I should note that my book **isn't** a "plan" either. I do include some specific recommendations (Chapter 18), but the primary purpose of this book is to break down each area in which progress needs to be made, see what tools are available to make that progress, identify the obstacles to making that progress and finally draw a conclusion about whether or not the necessary actions are likely to be taken. The "identifying obstacles" step will, hopefully, provide some guidance about where to apply resources to do what needs to be done.

My overall reaction to TIME's Special Climate Issue brings to mind the response of a movie critic who, in answer to a question about how he liked a particular movie, reportedly replied: "I liked everything about it except the concept and the execution." In this case, at least I liked the concept, but I had a lot of trouble with the execution. When I picked up the "Earth Survived" issue, I was skeptical, but I was hoping to learn that there WAS a path forward that would allow us to "beat" global warming. I was hoping there was something I hadn't thought about or didn't know.

I came away disappointed and frustrated. I came away feeling:

1. The assumption that Americans "will rise to the occasion" is unsupported wishful thinking,
2. The expectation that today's youth can solve the problem is misplaced and delusional,
3. The reliance on technological innovation to deliver a "green silver bullet" is unrealistic, and
4. TIME didn't really have a list of actions or a time frame for doing what needs to be done to support the premise that "Earth Survived."

I put down the magazine feeling that the average reader would surely conclude that we were making adequate progress and everything was going to be OK. If we just continue what we have been doing: putting solar panels on rooftops, erecting windmills in farmers' fields and offshore and researching the hell out of everything related to energy, then we can pass the keys for the EV to our kids and . . . Earth will be just fine.

"If wishes were horses . . ."

The Role of the Media: TIME's 2019 Climate Issue

18 WHAT WE NEED TO DO

"No pain, no gain."

Exercise Mantra, 1980s

What would you do if you were the only breadwinner for your family and you went to work one day and your employer told you that you were laid off? What if you knew that economic conditions didn't favor you getting a new job? Sadly, that's exactly what happened to literally millions of people, maybe even some of you, as the COVID-19 infection spread across the country and around the world in late 2019 and early 2020.

You'd probably start by taking a hard and discriminating look at all your expenses. You'd cut all of your discretionary spending: no more dinners out, no more movies, no more concerts, nada. Then you'd go deeper. Maybe you'd drop some of your cable services. That gym-membership would be history. You'd keep your phone, but you might go on a cheaper plan of some kind. You'd start using all the canned goods in your cupboard rather than buying new. When those ran out, you'd buy a lot more beans and rice and a lot less meat. If push-came-to-shove and if you had a garage, you might sell that old "whatever" that's been sitting in there for too long. You might try picking up part time work doing almost anything. For the first time in your life, maybe you'd visit your local food pantry to make a withdrawal rather than your usual contribution. You'd turn off the furnace or the air conditioner. You'd become fanatical about turning off lights. You might try to negotiate with your landlord or mortgage company. If all that wasn't enough, you might start using your credit card for necessities. The point is this: **you would view every expense in your life through a different lens.**

That's exactly what's needed in regard to global warming and climate change. We need to look at every (fossil fuel) consumption activity in our lives through a different lens. Global warming action requires adopting a mindset that **moves energy and emissions considerations to the forefront of everything we do.** Let me give you an example.

How would you react to a proposal that we ban the burning of natural gas for all decorative and commemorative purposes? This would mean, for example, extinguishing all

> *Extinguish all eternal flames everywhere and, blasphemy, permanently extinguish the Olympic torch.*

eternal flames, everywhere, and permanently extinguishing the Olympic torch. Will this save the planet? No, of course not, but it might serve as a visible reminder that we need to be thinking about energy and emissions issues in a whole new way. Perhaps a tiny, gold-plated sculpture that looked like a burning flame could replace the actual eternal flame at JFK's grave. Perhaps the Olympic cauldron could be replaced with red, orange, and yellow ribbons streaming upward, blown by a fan from below. If we went to those extremes, maybe people would be willing to ban those patio heaters that allow customers to dine in comfort, outside, on wintry nights. If nothing else, the debate that would be "ignited" over a proposal to ban such items would highlight the fundamental message of this book: **to even make a dent in CO_2 emissions, we have to start doing things radically differently than we have in the past. Every assumption we've ever made about energy issues will have to be open for reconsideration.**

While I argue with certainty that we're not going to "beat" global warming, I also know with the same degree of certainty that if we take certain actions **now**, we can slow the buildup of CO_2 in the atmosphere and that would be a good thing. In other words, the earth will be far better off if we're able to delay reaching an atmospheric concentration of 500 ppm CO_2 until sometime after 2075 rather than in 2050. Indeed, we'd be better off **never reaching** the 500-ppm level, but even with some dramatic emissions reduction efforts, that's exactly where we're headed.

The primary message of most climate change books that I've read seems to be, "this is how dreadful things could get, UNLESS we do x, y, z." The "unless" part of those books gives them a strangely optimistic twist. They seem to imply that we still have a chance to avoid the catastrophic futures these authors describe.

My message is different. My message is:
We haven't done x, y, z;
    we're not on track to do x, y, z;
        it will be expensive and disruptive to do x, y, z;
            powerful special interests will resist x, y, z; and
                when we finally get around to doing x, y, z,
it'll be far too little and far too late.

Put the "Big Rocks" In First

Most of you have probably heard the life-management story about filling a large glass jar with rocks, pebbles, sand, and water. If not, google "put the big rocks in first." It has a useful lesson even if most of us don't always abide by its principles.

In this chapter, I list some of the "big rocks" when it comes to combatting global warming – i.e., those things we're not already doing that are critical to changing the trajectory of our relentless march toward a much less pleasant climate future.

I've pulled this list from the Action Items I listed at the end of each chapter where each issue was introduced. **This list only deals with the larger policy and program initiatives. It is _not_ all inclusive and is intentionally limited to technologies and actions that are known and feasible TODAY.** Atmospheric GHG concentrations are increasing so fast, we have to begin immediately using our existing tools to the greatest extent possible to slash emissions. If something better comes out of the research labs in the near future, is perfected quickly and then deployed rapidly in huge numbers, great. But if a miracle cure doesn't materialize, at least we won't have wasted a couple (more) decades of on-the-ground progress cutting the sources of our CGI.

Many of the Action Items I list will require years to complete. When that's the case, there will need to be annual milestones so that we can stay focused on what needs to get done and know the extent to which we are, or are not, making progress toward our goals.

Most of these Action Items will also be implemented through national, state, and local governments. When authors like me urge their climate-action activist-readers to make their demands known to policy makers and corporate leaders, **these are the things that you need to _demand_ that your representatives do and support.**

1. Establish a worldwide carbon tax. Call it a "Planet Survival Investment" (PSI) charge so that individuals who consider taxes a "four-letter" word won't run for the hills as soon as they hear the suggestion. The tax must be universal and increase incrementally over time. There also needs to be an equivalent PSI charge on non-energy agricultural and industrial sources of GHGs.

2. Build two demonstration nuclear power plants (based on two different technologies) within seven years.

3. Begin reprocessing spent nuclear fuel (domestically or through an international agreement) within five years. Make a permanent nuclear waste disposal repository operational within 6 years. (Note: this action won't reduce GHGs, but it would remove an impediment that may be keeping a lot of people from supporting a technology that can (i.e., nuclear power).

4. Electrify all mainline and coast-to-coast train routes within 10 years. Move all long-haul truck traffic to the electrified railroads.

5. Increase mass transit capacity and efficiency. Make mass transit free.

6. Increase Corporate Average Fuel Efficiency (CAFE) standards (i.e., increase mpg) for cars and light trucks.

7. Identify and build large, pumped hydro-electric storage projects on an accelerated basis. Build more low-head, flow-of-the-river power generating plants.

8. Phase out all (remaining) coal-burning electric generating plants within five years.

9. Terminate all coal exports within five years.

10. Establish standards to dramatically reduce packaging waste. Establish product repairability standards. Require all durable products to be 100% recyclable.

11. Identify sites for and build micro-hydro and garbage-burning generating facilities on an accelerated time schedule.

12. Build water storage and delivery infrastructure. Build flood control infrastructure on an expedited basis. Ban garbage disposals.

13. Use levees sparingly and only to "buy time" to allow people and enterprises to depopulate areas prone to weather disasters.

14. End ethanol production within five years. Increase gasoline taxes by $0.30/year, every year, for at least the next 10 years.

15. Establish a National Emergency Response Corps, (NERC), a permanent, standing "army" of 10,000 firefighters and emergency response personnel who can be deployed around the country from one emergency to another. Equip NERC with an air force of fire-fighting aircraft and helicopters, (at least double the capacity we currently possess in the entire country). Equip NERC with other equipment to respond to floods, tornadoes, hurricanes, earthquakes and drought emergencies.

16. Require airports in cities greater than 500,000 to have mass transit links to their city centers. Revise zoning laws to encourage high-density, walkable living spaces.

17. End "cap and trade" programs. Until they're eliminated, make the cap-and-trade financial benefits for green products flow directly to the product owners and not the product manufacturers (i.e., EV owners, not EV manufacturers).

18. Increase taxes on water consumption.

19. Require landscaping and agricultural production to conform to regionally appropriate standards (e.g., no high-water-consumption crops in drought-prone regions).

20. Require product delivery services (Amazon, FedEx, UPS, etc.) to cooperate and coordinate the delivery of packages ("the last mile problem") most efficiently and with the least traffic impact.

21. Direct utility regulators to develop (modestly) lower electric system reliability standards. Develop enhanced strategies to accommodate power shortages. Implement dynamic energy pricing for consumers.

22. Privatize flood and all disaster insurances. After disasters, limit government aid to search and rescue activities. Get government out of the business of helping people rebuild in areas prone to weather-related damage.

23. Negotiate with all countries to establish their own population target goals. Provide aid to those who achieve goals that are consistent with lowering the rate of population increases. Support family planning to slow population growth.

24. Stop deforestation and the destruction of wetlands.

25. Increase EV licensing fees to be equivalent to gasoline taxes for road maintenance and to cover the extra expenses of special fire-fighting equipment for EV battery fires.

26. Adopt equal, dynamic, regional pricing for all energy suppliers. In other words, pay residential power producers (i.e., owners of

solar panel-systems) the same rate paid to commercial power producers at the time their energy is flowing into the grid.

27. Develop programs to rapidly increase energy conservation measures. Increase residential and commercial building insulation standards by 50%.

28. Terminate fusion power research funding. Terminate <u>manned</u> mission-to-Mars activities. Terminate all government-funded manned space missions beyond earth orbit. Reprioritize all budget allocations to maximize funding for planet survivability investments.

29. Institute a quarterly decarbonization report card tracking the atmospheric concentration of GHGs. Track and report the percentages of U.S. electricity consumption (retail and self-generation) 1) from all carbon-free sources, 2) from wind and solar generation alone, 3) from nuclear power, and 4) from energy storage.

30. Require all Environmental Impact Reports to include a consideration of GHGs.

The first item on this list, implementing a worldwide carbon tax, **would be the single most crucial step the world (and the U.S.) could take toward decarbonizing the world energy economy.** A carbon tax would finally include the externality associated with burning fossil fuels that's been ignored up until now, i.e., CO_2. It would also **immediately** make every one of the other items on this action item list moot or easier to implement. Consumers would automatically seek out more efficient cars and trucks, utilize mass transit more heavily, improve building insulation, be more receptive to nuclear power, invest in micro-hydro, low-head and run-of-the-river power plants, install solar panels, increase the number of solar panels in existing systems, carpool, buy more EVs, put up clotheslines for drying clothes, use less water, use fewer disposable plastic bags, demand less packaging waste, etc. It would also spur added private investment in energy R&D. Consumers and the marketplace would decide which projects and products would materialize and which would quietly disappear. The government would no longer pick winners and losers. Competition and the market would work their magic. And that market would be dynamic as the carbon tax inched up over time in response to the mitigation need.

As we think about what we need to do to slow the onslaught of global warming and how we'll respond to it, we'll

> **Carbon Tax:**
> *The projects and products that would materialize and those that would quietly disappear would be decided by competition in the market.*

need to avoid making assumptions that may not hold up well in that warmer world. For example, several of the actions I recommend make assumptions about precipitation patterns based on past experience. In recent years, drought has plagued the West and Southwest while flooding, especially springtime flooding, has ravaged parts of the Midwest. The recommendations to build infrastructure to move water from the Midwest to the thirsty West or to build more low-head or run-of-the-river hydroelectric power plants **assume** that past precipitation patterns will return; they probably won't. Some flood-prone areas may not have spare water to ship west in the future. As we plan our response to global warming and climate change, much of what we "know" about our world may have to be "unlearned."

In Chapter 4 I asserted that until we embraced the reality that we aren't going to "beat" global warming, **we were likely to waste (more) time, squander (more) resources on futile countermeasures and most importantly, reject solutions that could actually help mitigate the problem.** I also asserted that the way forward would require all of us to think differently about energy and examine all of the assumptions about energy production and usage that we hold dear.

Many of the measures in the above list will **not** be supported by a large segment of the population **unless** they feel some obligation to future generations and are willing to think differently about energy. Specifically, items # 1, 2, 3, 6, 7, 11, 13, 14, 17, 22, and 28 will almost certainly be unpopular with the general public (and perhaps you, personally).

The question is: if you're among those folks who have rejected these kinds of actions in the past, but desperately want to lessen the impact of global warming, **are you willing to reconsider your opposition to these proposals? Are you willing to think differently?**

Or, are you so opposed to these kinds of actions that you're willing to let global warming proceed to its full destructive

potential? Knowing we can't "beat" global warming, are you still unwilling to consider actions that could at least make the crisis less catastrophic for future generations? The choice is between doing something dramatically different or, effectively standing by while the earth continues to heat up. Which will it be?

I suspect (even hope) that this choice is the personal and moral dilemma that many reading this book, who may not have been "all in" on combatting global warming up until now, will now have to face.

In regard to countermeasures to address the consequences of global warming (for example, items #4, 7, 10, 15 and 22), wouldn't it be refreshing if our corporate entities and government representatives could do a better job of understanding what's coming and respond accordingly before the crises reach full blown proportions? In the case of recommendation #15, for example, (creating a National Emergency Response Corp) what if some underpaid intern on an elected representative's staff read these words and convinced his or her boss, the Representative or Senator, that creating the NERC was a good idea? How many lives might be saved? How much property damage could be avoided? How much misery might never become two-inch banner headlines in the local paper?

(Competitive) Advantage to the Efficient

Industry and the private sector could wait for regulators to dictate stricter standards for their products, **OR** they could take the initiative and claim a competitive advantage for themselves and their products by building greener. Consumers can, in turn, influence the marketplace by gravitating toward the more durable, more repairable, more efficient products, even if their initial cost is higher.

Manufacturers of refrigerators, for example, could market a programmable model with built-in "cold storage" so the refrigerator/freezer units could be programmed to maintain the desired internal temperatures for fourteen to sixteen hours without running the compressor. Customers could program their refrigerators to remain "off" at night and turn "on" during daylight hours when solar panel would provide the necessary power. Other customers might program their refrigerator to access super off-peak electric rates from their local utility.

If hot water heater tank manufacturers want to continue producing natural gas fired units, they could at least include an access port into which an electric heater element could be inserted at a future date to convert the unit from gas to electric after the grid's gone green.

Think Globally, Act Locally, (Implement Personally)
While these policy issues are most assuredly (some of the) "big rocks" in terms of slowing the buildup of CO_2 in the atmosphere, for most of us, as private citizens, complex policy initiatives often seem to evolve slowly and feel remote. Besides voting, demonstrating and pressuring public officials, fortunately, there are other things we can all do in our own lives to lighten our own personal carbon footprints. Book after book after book lists the things all of us can do. These are actions you've heard about for years: set your thermostat higher in summer and lower in winter, carpool, add insulation, consume less meat, consume less of everything, landscape with drought-tolerant plants, shower with a friend, drink tap water, not bottled water, etc.

At the heart of all these lifestyle choices is a simple concept: consume responsibly, consume less. No doubt you've seen those neighborhood signs urging drivers to: "Drive Like YOUR Kids Lived Here." The global warming equivalent is: "Consume Like YOU'LL Be Living Here 100 Years from Now."

> *Consume Like YOU'LL Be Living Here 100 Years from Now.*

I could list all the quirky little things I do to minimize my energy and material consumption. In fact, in an earlier draft of the book I did just that. But I decided that that would be like serving you a fish dinner when instead, I'd much rather give you a fishing pole.

My suggestion is this: make a commitment to go through an entire day, one day, doing everything you normally do, but with **energy foremost in your mind at all times**. Throughout the day, repeatedly ask yourself, "Could I use less energy to do what I'm doing?" Could I use fewer raw materials (including water)? How could I lessen the environmental impact of what I'm doing? In a sixteen-hour day, you might ask yourself those kinds of questions every five minutes (roughly 200 times in the course of the day) and in the process, generate a list of ways you can lighten your own personal carbon footprint.

The list you come up with will be far more meaningful **AND FAR MORE ACTIONABLE** for you personally than any list I might suggest. After doing it for a day, expand to a week and begin acting on your list. Remember the whole point of the exercise is to bring energy considerations to "front-of-mind" in everything you do.

Think about it this way: suppose someone gave you $1,000 when you woke up tomorrow morning and told you that you could keep that fraction of the money that was equivalent to the fraction of the energy that you DIDN'T use in the course of the day compared to what you normally use. In this make-believe exercise and in real life, you'd have skin in the game.

In addition to the above, recycle everything and when you're not recycling, conserve, conserve, conserve.

Friendly Fire

Three books with advice about what we can do about global warming have come out fairly recently. One is by actor and activist Jane Fonda: What Can I Do? My Path from Climate Despair to Action. The second is by Bill Gates: How to Avoid a Climate Disaster: The Solutions We Have and the Breakthroughs We Need. Another one is Drawdown: The Most Comprehensive Plan Ever Proposed to Reverse Global Warming, edited by Paul Hawken. Each takes a unique approach to the issue and adds something of value to the conversation. I encourage you to read the first two and at least browse the third; it's pretty dense.

All three reach the conclusion that we can "beat" global warming which, of course, is contrary to the unhappy conclusion I've reported in this book. Accordingly, I'd like to make some brief observations about each. Like my review of the 2019 Special Climate Issue by TIME magazine (Chapter 17), I feel a little uncomfortable criticizing my fellow authors who've put their ideas down on paper. I know they each care passionately, as do I, about how the climate story unfolds long after we're gone. I wholeheartedly support their efforts to find ways to do something about global warming but disagree with their conclusion that we will "beat" it. This is important because the different conclusions each of us reach, give rise to a different set of actions.

Jane Fonda is an activist, so it's not surprising that she comes at the issue of global warming from an activist's perspective – raising awareness, bringing attention to the issues, getting arrested and recruiting other people to do the same. Her book is a narrative describing the Fire Drill Friday demonstrations she and a number of her high-profile friends are conducting in and around Washington D.C.

She and her compatriots have rightly identified that climate change will have the greatest impact on the less well-off and have, consequently, married social justice objectives with climate change action. It would seem like a marriage made in heaven since those who are most committed to combating global warming are also more likely to support social justice initiatives and vice versa. If she can advance both causes, more power to her. I agree with most of what Ms. Fonda is saying about societal issues (income inequality, the concentration of wealth, job security, childcare, health care, etc.) but it sometimes felt like the links between global warming and social justice got lost in "feel-good language." I'm an engineer; I need concrete suggestions, actionable plans.

In a chapter on Climate, Migration and Human Rights, for example, one of the speakers at the information rally was bemoaning as "unconscionable" the action by PG&E to de-energize electric circuits in regions where high winds could cause power lines to spark forest fires. These actions were taken by the utility in the aftermath of the Camp Fire in the prior year that was determined to have been sparked by power lines. The fire obliterated the town of Paradise, CA, killed 85 people and literally bankrupted the utility. The speaker proclaimed: "The resiliency plan should become a movement that ends up with the people owning the energy they need."

Only a portion of the speaker's words were quoted in the book, so perhaps there was more of an explanation in the full speech that followed, but try as I might, I have no idea what the speaker was trying to say. That's what I mean when I say that the global warming and social justice issues seen to get muddled in feel-good language.

The problem in these remote, forested communities wasn't an issue of who "owned" the energy [resources], it was a problem related to transmission lines and distribution wires in adverse weather conditions. Was the speaker advocating undergrounding the transmission and distribution wires? Was he advocating eliminating all power lines and disconnecting from the grid entirely? Even the most hard-core renewable power advocates who aren't illegally cultivating marijuana plants in remote locations somewhere, maintain a connection to the grid if they can. The town and/or individuals could disconnect from the grid completely by installing solar panels and banks of batteries, but then they'd have to cope with a frequency of outages that would have the residents clamoring to be reconnected. And besides, the attraction for most who live in heavily forested areas are the trees and those areas aren't typically the most favorable locations for residential solar panels.

Ms. Fonda's book is a call to action, but, from my perspective, it's often unclear about what those actions should be. When they are specific, it's unpersuasive that the called-for actions would "beat" global warming.

Ms. Fonda's book is silent on the issue of nuclear energy. She starred in the film The China Syndrome which debuted in theaters just twelve days before the Three Mile Island nuclear accident. I presume (but I don't know for sure) that she was not in favor of expanding the use of nuclear power in the U.S. at the time. That was more than four decades ago and back then only a handful of scientists had even uttered the words "global warming" or "climate change."

Now that the science of global warming is better understood and its effects are real, observable, upon us today and likely to become much more unpleasant in the near future, I wonder if her views on carbon-free nuclear energy have changed? I wonder if one of the things on her list of "Things I Can Do" includes becoming more knowledgeable about the pros and cons of the newer, passively safe reactor designs? I wonder if she's familiar with the work of environmentalist James Lovelock who made the transition from antinuclear to pro-nuclear advocate as one of the steps to combat global warming. She may want to write an addendum to her book. If so, I have a suggested title: What Can I Do? My Path From Antinuclear Activist to Supporting the Only Base-Load Generation Technology That Can Get Us Out of the Global Warming Mess We're In. But seriously, in the context of our CGI, her reconsideration of the expanded role that could be played by nuclear power would be a notable example of thinking differently about the issue once the realities of meaningfully combatting global warming are fully understood. **If a climate activist of Ms. Fonda's passion, visibility and influence re-examined the issues (as Dr. Lovelock has) and came out in favor of the modern, passively safe nuclear options that are available today, she could truly move a lot of people from despair to action.**

The second book is by Bill Gates. Gates is a self-described technology nerd and that's how he approaches the climate question

He breaks down the global warming problem into its constituent parts, surveys where we are and what technological breakthroughs we need.

I agree with most of what Mr. Gates has written, but we're at polar opposites on at least a few of his perspectives.

Like me, Gates is an unabashed supporter of all renewable energy technologies **and** like me, he has simultaneously identified nuclear power as the critical technology necessary to slow global warming. He states that he's personally invested a billion dollars in energy technology development and although he doesn't say so specifically, it appears as though the single largest investment in his personal energy R&D portfolio is in nuclear technologies. In his book, Gates writes:

> *"Here's the one-sentence case for nuclear power: It's the only carbon-free energy source that can reliably deliver power day and night, through every season, almost anywhere on earth that has been proven to work on a large scale." "No other clean energy source even comes close to what nuclear already provides today."*

After releasing his book, Mr. Gates engaged in a media blitz to promote it. Being one of the wealthiest and best-known individuals on the planet and perhaps the most generous philanthropist, Mr. Gates had no shortage of interview opportunities. He is, after all, Bill Gates, right?

I listened to several of his interviews and had previously watched the Netflix series, Inside Bill's Brain: Decoding Bill Gates, with a particular interest in Episode 3 in which his nuclear technology activities were highlighted. Given Gates' understanding of and investment in nuclear technology (both dollars and passion) and the role it could play in helping to change the progression of our climate crisis, I was astounded that **not once in all the interviews I heard or watched did an interviewer bring up the issue of nuclear energy and not once did Gates find a way to inject the issue into the discussion.**

I have my suspicions that this may not have been an oversight. Gates was promoting his book and it's no secret that nuclear power is a technology that often elicits strong negative responses. Perhaps Mr. Gates didn't want to wade into those waters for fear the nuclear discussion would swamp the bigger message he was trying to promote.

I don't know if that was the reason, but I just wish this had been a case where Mr. Gates was willing to **put his mouth where his money is**.

Gates does an excellent job of laying out the CO_2 problem and what needs to be done. He also says explicitly (and many times), "this is going to be hard." The discerning reader will recognize that what he's really saying is "this is going to be expensive."

Unfortunately, the solution he proposes, his "detailed plan," is unacceptably thin in terms of details. The central element in his plan to "get to zero" is to rely on quantum leaps in innovative technologies that will flow out of a five-fold increase in energy R&D. The new funding would be ramped up over 10 years and would include high-risk, high-reward projects. Unfortunately, policy makers and legislators don't appear to have read his book; no such level of new R&D funding has materialized or even been proposed.

Mr. Gates also seems to reject the incremental reductions in CO_2 emissions that could be realized by deploying technologies we have in hand today. This is one area where we differ. He uses the example of replacing a coal-fired generating station with one burning natural gas. Doing so, he explains would reduce CO_2 emissions today but would make it harder to reach zero CO_2 by 2050 because, once built, there would be a strong incentive to continue operating the gas-fired generating station well past 2050 (i.e., for its full 40- or 50-year expected lifetime) to recoup the original investment. He's correct to point out that incremental improvements (in any technology) make it harder for new technologies to penetrate the market. Examples of this phenomenon are everywhere: high mileage gasoline-powered vehicles diminish the benefits of an EV, insulating your home may diminish the benefits of going solar, installing low flow flush toilets may diminish the incentive to transition to waterless toilets and so on.

But it's a risky proposition to assume that game-changing breakthroughs will emerge from research that hasn't yet been funded. We can't wait. We have to use every tool in our toolbox to reduce emissions TODAY without waiting. Besides being risky, betting on technological breakthroughs may reflect a kind of technological arrogance that's especially inappropriate given the past five or six decades during which top-notch scientists and engineers around the country and around the world have been looking for exactly the same kinds of technologies he's identified that we still need today. In any case, if a better technology solution were to come out of the lab and that technology was head and shoulders better than anything else, the transition to that better technology wouldn't be appreciably delayed by earlier investments. That's why economists developed the concept of stranded investments and tax codes include provisions for accelerated depreciation and write-downs.

While I am heartened by Mr. Gates' support for advanced nuclear power plants, I don't necessarily agree with everything his team is proposing. At the risk of allowing this discussion to stray into "wonky" territory, the Gates team has inexplicably included a molten salt energy storage subsystem in its nuclear plant design which makes little or no sense for a unit with a low marginal operating cost. **They should make the reactor and all of its auxiliary systems as simple and as robust as possible.** We can add the bells and whistles later. At this stage, all that really matters is that the new plants can be built and licensed in a reasonable period of time, at a predictable cost and can be operated safely. I would also urge his team to be ruthlessly honest in all their communications. They would do well to adopt the mantra: "under promise and over deliver." That's the only way the nuclear option will regain the public's trust and support.

I feel pretty certain that if Jane Fonda had come out in favor of nuclear power, we'd all know about it, so I'll assume for the moment that she's opposed. Mr. Gates is strongly in favor of renewables and advanced nuclear technology. They are both equally passionate that global warming must be thwarted. I would love to hear the two of them engage on the issue, superstar-to-superstar. I'd gladly moderate that conversation. Bill, Jane, call me?

Mr. Gates concludes his book writing this: "I'm an optimist because I know what *technology* can accomplish and because I know what *people* can accomplish." "We can keep the climate bearable for everyone, help hundreds of millions of poor people make the most of their lives and preserve the planet for generations to come." – truly an inspirational and aspirational conclusion.

I too am an optimist, but I also know that technology has its limits. People also have limits, not only in terms of what they can develop but also what they're willing to accept, support and endure. My conclusion that we will not "beat" global warming is not fundamentally a conclusion about technology. It's a conclusion based on the interplay of technology, human behavior, international relations, economics, the time frame required to effect change, the magnitude of the challenge, the changing nature of the problem and a recognition of the potential for unknowns to thwart our best laid plans.

My message may not be as inspirational or aspirational as that of Bill Gates, but mine reflects the real world we live in.

The third book is Drawdown, edited by Paul Hawken. This is a dense compendium of technologies and areas of research relating to energy, buildings, mobility and especially food production and land use, all examined for their potential to reduce GHG emissions. From my perspective it's more of a reference document (a menu), than a plan. It ranks activities in terms of their potential for emissions reductions or reversal. It doesn't focus on a schedule for the utilization of the various green technologies it discusses or the progression of emissions reductions over time. It does a good job of summing emissions reductions potential, but emissions reduction potentials are different from emissions reductions.

The book includes a section dealing with women and girls. It notes that the ravages of global warming are likely to fall disproportionately on the less educated because they typically have fewer opportunities to respond in times of economic stress. Women, the book explains, will be particularly disadvantaged because in many parts of the world, education is often seen as less critical for women.

In addition, the book highlights the real potential of how educating women can help reduce emissions. This is in part because educated women are central to so much economic activity especially in less developed countries and because educated women have fewer children. The article cites a Brookings Institution study that claims, "The difference between a woman with no years of schooling and [a woman] with twelve years of schooling is almost four to five children per woman." I whole-heartedly endorse smaller average family sizes and equal educational opportunities for all youth, boys, and girls alike, but the article ignores the added greenhouse gas emissions that will result as more highly educated people, men and women, require and demand more goods and services. We should strive to improve the lives of the world's less fortunate, but if we're successful, we'll need to find ways to reduce other emissions even more to compensate for the added emissions that accompany improvements in the standard of living of our fellow citizens of the world.

The compendium also appears to have a distinct anti-nuclear bias. The section on nuclear power (3 pages out of 200) includes an editor's note:

"One hundred solutions are featured in <u>Drawdown</u>. Of those, almost all are no-regrets solutions society would want to pursue regardless of their carbon impact because they have many beneficial social, environmental, and economic effects. Nuclear is a regrets solution . . ."

Compare this perspective on nuclear power with the one paragraph perspective by Bill Gates that I quoted a few pages ago.

As you think through the Gates perspective and the Hawken perspective, recall that each year for the last twenty years, nuclear power has been the carbon-free generating resource for between seventy and seventy-five percent of all the electricity consumed in France.

Without regrets, I offer the opinion that perhaps one of these two perspectives on nuclear power might just be, (let's see, how can I put this tactfully?) . . . wrong.

19 SCENARIOS AND SCORECARDS

"Never let a serious crisis go to waste."

Popularized by Rahm Emanuel (and others) in the aftermath of the financial collapse, 2008.

Transforming our worldwide, fossil-fuel-based economy to an economy powered by renewables and other carbon-free energy sources will be expensive, difficult, and slow. The pace of change is picking up, but it's not happening at a rate anywhere near quickly enough to avoid the worst consequences of our catastrophic global imbalance.

Critical Factors

What are the things that would have to happen for the "world" to step up and confront the global warming challenge in a meaningful way (i.e., to do the things I described in Chapter 4)?

First, the nations of the world would have to reach some kind of consensus (Chapter 14) that the problem needs to be confronted and make commitments to do what needs to be done. Then, each nation would have to identify the technologies and processes most applicable to their circumstances to achieve the necessary emissions reductions. And finally, there would need to be some kind of a mechanism to ensure that the agreed-upon plans were actually carried out.

In the past decade, many steps in the right direction have been taken. In the coming years, more progress will be made, but not enough. Depending upon what actions are agreed to and how quickly they're taken, a range of possible scenarios for the world's energy future can be envisioned.

With a focus on the U.S., Exhibit 19.1 lists some (not all) of the critical factors that will influence those future scenarios. Each scenario depends on which of the actions in Exhibit 19.1 (shown with an "X,") are taken. Scenario #1, for example, is a continuation of what's happening today, the status quo, consisting of Actions #1, 4, and 8. It's the least aggressive response to global warming.

Table 19.1 Critical Factors Influencing Future Scenarios

Action		Scenario					
		1	2	3	4	5	6
1.	Continued Deployment of Renewables on a Cost/Benefit Basis	x					
2.	Accelerated Deployment of Renewables		X	x	x	x	X
3.	Significant Quantities of Bulk Electric Energy Storage				x	x	X
4.	Ramping Down Nuclear and Hydroelectric Capacity	x	X	x	x		
5.	Ramping Up Nuclear and Hydroelectric Capacity					x	X
6.	Lifestyle Adjustments to Move Energy Demand to the Daytime				x	x	X
7.	Aggressive Dynamic Energy Pricing		X	x	x	x	X
8.	Gradual Phase-In of EVs	x	X		x		
9.	Accelerated Phase-In of EVs & Phase-Out of Gasoline Cars and Light Trucks			x		x	X
10.	A Consensus to Act Aggressively					x	X
11.	Train Electrification. Biofuels or Syn-Fuels for Jets, Heavy Trucks, Etc.					x	X
12.	International Cooperation Including a Worldwide Carbon Tax						X
13.	Countries Promote Lower Birth Rates						X

With public sentiment to "do something" becoming ever stronger, Scenario #1 is too pessimistic in terms of what's likely to be done to combat global warming.

Scenario #2 reflects growing public pressure to do more than we're doing today. It incorporates some of the actions that nations are taking already or will take in the near future to move in the right direction. It assumes we take Actions #2, 4, 7 and 8. These are all things individual nations can do on their own.

Scenarios #3, #4 and #5 are scenarios in which progressively more aggressive actions are taken.

Scenario #6 is the most aggressive, requires international cooperation and includes Actions #2, 3, 5, 6, 9, 10, 11, 12 and 13.

Let me offer two narratives using the same approach Bill McKibben used in his essay in TIME's Special Climate Issue. I'll look back from the year 2050 at how we responded to a year of extreme weather events. First, I'll describe the fictitious events of a catastrophic year. Then I'll lay out two different responses, one where we implement the actions described as Scenario #2 and the other where we implement the actions in Scenario #6.

First, a description of the extreme weather events.

A Mind-Focusing Year

Global warming continued to be a political hot potato in the early 2020s. Demonstrations by young activists received a lot of press coverage and helped raise awareness and concern. More than 60% of Americans agreed that global warming was real and caused by human activity. Support for measures to combat global warming was increasing steadily, but no clear consensus developed until . . .

On September 19, 2023, Nila, a slow-moving Category 5 hurricane, came ashore between Miami and Fort Lauderdale at high tide. The storm surge inundated both cities. Nila tracked northwest across the state, through Tampa and into the Gulf of Mexico. The now-Category 4 storm regained strength over the warm Gulf waters and took a turn to the northeast. Nila was again a Category 5 storm when it slammed into the Florida panhandle and began a slow, relentless, and destructive path across southern Alabama, through Georgia and up the Eastern seaboard. By the time the storm reached Virginia and Maryland, it had been downgraded to a tropical depression. The storm lingered over Hagerstown, Maryland where a record 21 ¼ inches of rain fell in a 48-hour period contributing to massive flooding. Mercifully, the storm finally tracked out into the Atlantic after causing more flooding in eastern Pennsylvania and New Jersey.

To make matters worse, Nila spawned over forty major tornadoes in eastern Kentucky and the Ohio River valley as it churned up the East Coast. The worst of the tornadoes took a direct hit on the city of Louisville, the first time an E-5 tornado had struck a major metropolitan center.

"Old timers" compared Nila to Superstorm Sandy, but Nila was actually much worse. Over the course of its six-day reign of terror, Nila affected 47 million people, was responsible for 297 deaths and caused an estimated 420 billion dollars in damage, making it many times more destructive than any storm in American history. Each of the ten states along the path of the storm sustained more property damage and loss of life from Nila than each had experienced from the worst storm in each state's history.

While the East Coast was drowning in too much water, the record-breaking drought in the West continued unabated. The West and Northwest experienced their worst fire seasons in recorded history. Three of the 1,807 major blazes were more than noteworthy. The Hayden fire in northwest Idaho burned an area the half the size of Rhode Island and obliterated dozens of mountain communities. Fortunately, the fire burned away from and spared Coeur d'Alene. The Lassen fire in northeast California burned an area only slightly smaller than the Hayden fire but was more deadly and destructive. The granddaddy of them all in terms of sheer size, however, was the Yukon fire that charred an area half the size of Maine in a mostly unpopulated inland area of southeast Alaska and northwest Canada.

But it wasn't just the coasts that felt nature's fury. Record flooding up and down the Mississippi in the spring preceded an historic dry spell in late summer. Dry thunderstorms and high winds triggered raging fires that swept across enormous swaths of farmland in eastern Colorado, Kansas, and Nebraska. Kansas got the worst of it. The fire incinerated fully 18% of farm production along with countless farm buildings and homes in Kansas.

And now, looking back from the year 2050, how might we have responded to the catastrophic year of extreme weather in the early 2020s?

Scenario #2 – The U.S. and the World "Snooze"
After Nila, the nation declared a 3-day period of mourning. Congress authorized the president to redirect funds from every sector

of the budget to pay for relief and rebuilding. The long standing and controversial subsidies usually paid to the fossil-fuel industries were among the many programs wiped off the books. Climate change activists hailed the move, but coal and oil industry lobbyists were ultimately able to quietly secure tax credits in lieu of the direct payment subsidies.

In the aftermath of the storm, the focus of the country was on rebuilding and reenergizing the economy in the devastated areas. Impassioned appeals to finally take bold action to combat climate change were characterized as unpatriotic at a time when so many Americans needed assistance. Individual states continued to offer rebates to promote the deployment of solar panels and electric vehicles, but even Nila and the worst fire season in history couldn't move the nation to aggressive climate action.

As the economy gradually recovered, Congress narrowly passed a national program to promote renewables and EVs. Nuclear power advocates succeeded in adding provisions to encourage utilities to build new nuclear plants. No utilities were able to utilize the assistance, however, because they all anticipated a surge in solar power resources. Additionally, the public utility commissions in most states were paralyzed by a vocal minority opposing nuclear power. In the late 2020s and 2030s, some nuclear plants received license extensions allowing them to keep operating, but others were shut down. By 2040, the contribution of nuclear power to the national grid in the U.S. declined from 17% to just over 8%. New reactor concepts were well along in development, but concerns over safety and long-term waste disposal caused acceptance by the public to lag far behind.

With Federal subsidies, solar and wind generation resources began to appear not only in solar-rich and wind-rich areas, but more broadly across the country. Critics complained that the subsidies were too large. Wind turbines proved to be the more favorable of the renewable assets. Although intermittent in nature, wind resources were far more dependable because they were able to generate electricity at all hours of the day. One utility executive observed, "The wind is nearly always blowing somewhere."

Solar power, although nominally larger in installed capacity than wind resources, didn't enjoy the same success. Utilities were caught between a rock (trying to keep electric rates from soaring) and a hard place (struggling to maintain system reliability). Virtually all utilities had to continue operating their fossil-fuel power plants which, by this time were predominantly using natural gas. All but two domestic coal burning generating stations had been converted to natural gas by 2038. The fossil-fuel-fired power plants, however, served not only as backup resources to be ramped up on overcast days when solar output dropped, they continued to be the backbone of the U.S. grid by providing 75% of all electric energy in the overnight period from sundown to sunrise. The capacity factor of the fossil-fuel units (i.e., the actual output of these units compared to what they were capable of generating if operated at full power continuously) remained steady at 45%, but their heat rate (a measure of how much energy is required to produce a unit of electrical energy) inched upward. Nationally, the heat rate increased by 9.7% which means that the CO_2 emissions per kWh from fossil-fuel-powered generating stations were the highest they had been in 70 years.

Utilities complained that federal subsidies for energy storage were more important than new solar panels, but those complaints were ignored. Batteries weren't as exciting or as visible as solar panels which remained the darling renewable of politicians.

After utilities implemented dynamic pricing (whereby energy prices change continuously depending on the availability of supplies and wholesale energy costs), the behavior of American households did show some changes. Automated households managed their energy usage based on energy costs. Households with less automation relied on the occupants to turn off lights and turn thermostats up in summer and down in winter.

EVs were also heavily subsidized, much to the benefit of the mainstream auto manufacturers. As the major automakers entered the market and carbon credit programs were phased out, Tesla, the early leader in the field, shrank to become a boutique brand. EVs were no longer a novelty; they were everywhere. By 2050 EV sales grew to nearly 39% of all auto sales in the U.S. (mostly along the coasts, and in the South where their demand had always been strong). EVs rose to 26% of all vehicles on the road in the U.S. The comparable fraction in the EU was nearly 52%, in no small part because gasoline taxes in the EU increased to nearly $7 per gallon. Gasoline taxes in the U.S. also increased from around $0.26 per gallon to just over $1.15. In an effort to preserve market share, oil companies continued to drop wholesale gasoline prices to offset the higher federal taxes.

Unfortunately, the massive conversion to EVs didn't reduce transportation sector emissions because the marginal fuel used to generate the electricity to recharge EV batteries continued to be natural gas.

With the forest fire season extending over much of the year in the West and more severe storms especially in the East, it seemed as though the federal government was always playing catch-up. It could never quite make the investments to reduce CO_2 emissions because it was spending so much money helping communities recover from severe weather events. Levees and seawalls were built, but those defenses were often no match for Mother Nature. The nation finally raised the white flag and bought out the landowners in a few isolated communities until the price tag became so large that the practice had to be suspended.

Inaction by the U.S. made it virtually impossible for other, more-committed nations to ask their citizens to make meaningful sacrifices. European countries that were undertaking aggressive programs to cut emissions (and succeeding) tried to organize a trade campaign against the U.S. They proposed an environmental tariff on all U.S. goods both to punish the U.S. for not reducing emissions and to level the playing field so products produced in their own countries wouldn't be undercut by cheaper imports from nations not taking aggressive action to reduce emissions. In the end, the market power of the U.S. proved too much for the fledgling movement. The U.S. lagged well behind most of the developed nations and became something of an outcast in international circles. Polls showed a slow but steady decline in how the U.S. was perceived around the world as the effects of global warming worsened.

Severe storms and summer heat waves continued to take their toll in lives and property damage around the world. In affluent countries, the losses were measured in dollars; in poor countries, the losses were measured in corpses. Scientists continued to sound the alarm as the concentration of CO_2 in the atmosphere approached 500 ppm, almost 80% higher than what it had been in the pre-industrial age.

The health of the oceans continued to degrade over the nearly three decades after Nila: ocean acidity continued to increase, sea levels continued to rise (mostly due to thermal expansion initially), fish populations continued their downward trajectory and land-based ice reserves continued to shrink. Water-rights and famine-related conflicts escalated around the world. High-temperature records continued to exceed those established in the prior year almost everywhere.

Most nations felt compelled to take care of their own first which put a strain on international relations. Efforts to establish an equitable and effective world strategy to combat global warming continued. Conferences, both before 2023 (such as COP26) and after, always seemed to end with hopeful official statements, but no meaningful action plan. After much negotiation, a worldwide, incrementally increasing carbon tax was finally adopted with much fanfare in 2035. The initial price was set at a level to encourage a migration away from fossil fuels but also allow time for all parties (governments, businesses and individuals) to change directions. Over time the carbon tax was slated to become larger, but still didn't have the bite necessary to significantly alter the rate of emissions. Nations were to use the proceeds from the carbon tax within their own borders to accelerate the movement away from fossil fuels. Instead, some nations simply reimbursed their citizens for increased energy costs. Nations were also required to police themselves, but there were accusations and evidence of widespread non-compliance.

At international conferences, discussions (i.e., fierce arguments) continued about whether or not to utilize geoengineering measures (specifically injecting SO_2 aerosols into the upper atmosphere) in a last-ditch effort to slow the rate of warming.

Scenario #6 – The World Wakes Up

After Nila, the nation declared a 3-day period of mourning. Congress authorized the president to redirect funds from every sector of the budget to pay for relief and rebuilding. The controversial subsidies paid to the fossil-fuel industries were among the many programs wiped off the books. Coal and oil industry lobbyists tried unsuccessfully to secure tax credits in lieu of the direct payment subsidies and suffered a backlash of public resentment for even trying.

In the aftermath of Nila, political leaders in the U.S., sensing that the nation was finally making the connection between severe weather events and global warming, seized the opportunity to act. With the support of both houses of Congress, the President made a speech to the United Nations General Assembly pledging that the U.S. would unilaterally and dramatically reduce its domestic CO_2 emissions within a decade. He also announced that within two years, the U.S.

would only align itself with nations taking aggressive actions to address global warming. Until a global carbon tax was implemented, he said the U.S. would levy climate taxes on goods imported to the U.S. from nations failing to adopt aggressive GHG reductions. Among other things, the government prevented Americans from traveling to nations that weren't meeting emissions reduction targets. Foreign aid was conditioned upon participation in climate stabilization efforts. Climate activists hailed this change of direction by the nation with the highest per capita CO_2 emissions, but remained skeptical until, at the end of the first year, Congress passed a budget that explicitly incorporated all of the president's programs.

Nations that had undertaken aggressive programs to cut emissions, reduce family sizes and protect natural lands were rewarded with economic aid. Countries that had previously expressed reservations began to find ways to comply. Holdouts who remained committed to fossil fuels saw foreign investment, tourism, trade, and general economic activity begin to shrink. American consumers, responsible for so much of the excess CO_2 in the atmosphere, were finally wielding their collective buying power for environmental good.

At home, the president didn't sugar-coat the fact that Americans would have to make the biggest changes and the biggest sacrifices. The voices of opposition were drowned out by a two-word mantra: Remember Nila.

Domestic manufacturing to produce goods that might otherwise have come from foreign suppliers began to ramp up. Tariffs collected on imported goods from non-complying nations were funneled back to support domestic, carbon-free manufacturing. Federal programs to accelerate the deployment of energy storage facilities, renewable generating resources and electric vehicles were enacted. Gasoline taxes were increased by $1.00/gallon immediately and increased $0.21/year every year for the next six years. Existing internal combustion engine powered cars dropped in value. EVs were favored in the marketplace. The federal government enacted progressively more stringent building and appliance efficiency standards.

In time, the unilateral action by the United States evolved into a more robust version of the Paris Climate Accords. The United States, for its part, accepted its responsibility as the economy that had pumped the most CO_2 (per capita) into the atmosphere. Already on its way to dramatically cutting CO_2 emissions, the U.S. agreed to reduce its per capita CO_2 emissions to the increasingly lower levels that were being achieved in Western Europe. The U.S. also became the major funder of programs to resettle the populations of island nation being swallowed up by rising sea levels. China agreed to reduce its per capita CO_2 emissions to 30% of those in the EU and accelerated its construction of civilian nuclear power plants. One large coal-fired generating station near Xi'an was shut down permanently, mostly to improve air quality in the city, but yielding a secondary benefit of reducing CO_2 emissions.

The world's largest economies collectively agreed to financially isolate all governments that refused to join the worldwide effort to combat global warming. In a very short time, every nation signed on and agreed to stringent inspection protocols to verify compliance. After much negotiation, a worldwide, progressive, and incrementally increasing carbon tax was finally adopted. The climate tariffs initiated by the U.S. were discontinued and replaced by the negotiated world carbon tax. The initial tax was set at a level to encourage a migration away from fossil fuels but also allow a limited amount of time for all parties (governments, businesses, and individuals) to change direction. In time, the tax increased sufficiently to dramatically alter consumer choices. Clean energy capital investments, nuclear reactor development and a UN-sponsored rewards program for countries meeting their climate change goals were all funded by revenues from the carbon tax. Economic migration decreased and repatriation of former climate migrants to their homelands increased as regional job opportunities grew in nations actively engaged in thwarting climate change.

In the U.S., a public-private partnership to deploy advanced nuclear power plants was developed. Two advanced nuclear reactor concepts were selected, designed, and underwent component testing. Both prototypes were built on an accelerated schedule in the early 2030s and performed as expected. The first new, commercial, advanced nuclear plant went online in 2039 and was quickly followed by clones around the country over the next decade. By 2050 wind, solar, nuclear, hydroelectric, and biomass-burning power stations powered 64% of the U.S. power grid. With current commitments, that number will grow to 87% by 2075, driven mostly by new nuclear capacity. As the grid became less CO_2 intensive, the source-based net emissions from EVs began to drop. The transition away from natural gas burning in the residential, commercial, and industrial sectors accelerated as the carbon tax began to bite deeper. Legislation was passed to phase out new gas water heaters, clothes dryers, heaters, and stoves over a five-year period. As the existing stock of these gas appliances reached the end of their expected lifetimes, they were gradually replaced with their electric counterparts.

The massive quantity of electrochemical (battery) storage capacity deployed early in the renewables program was repurposed to add stability to regional power grids. The expanded high voltage transmission network across the nation continued to move massive amounts of energy regionally to maximize the utilization of carbon-free generating resources.

There were also massive investments in public transportation. Using subsidies derived from carbon tax funds, several regional systems made mass transit ridership free. Bike paths were expanded and bike usage, which exploded during the 2020-2022 pandemic, continued even as the economic and medical consequences of the virus gradually faded away. High-speed, inter-city passenger rail travel (e.g., NYC – DC, NYC – Boston, DC – Miami, LA – Las Vegas, LA – SD, LA – SF, Seattle – Portland, Chicago – DC, Chicago – St. Louis, Dallas – Houston, Houston – New Orleans) expanded beyond all expectations. Subsidized passenger rail service was half the cost of air travel and in many cases, faster (downtown-to-downtown).

Passenger rail service in the U.S. was still not on a par with that in the EU, but freight rail service was as good or better. The heavy railway network was nationalized and electrified from coast-to-coast over a two-decade period. The ton-miles of freight moved by rail increased by a factor of three (even as coal tonnage dwindled to near zero). Long haul trucking virtually ended as rail transport was cheaper and faster. Nationwide changes in land use policy were enacted to promote the development of more compact, high-density, people-friendly population centers.

Some of the areas in Florida and up the Atlantic coast that had been devastated by Nila were never rebuilt. The National Flood Insurance Program was privatized. Insurance premiums for flood-prone properties increased rapidly over a 20-year period. As the cost of insurance in these vulnerable areas grew ever-larger, owners jumped at the chance to have the federal government buy up their properties and turn them into recreational areas while they were still above the high-water line.

As energy prices increased because of the carbon tax, consumers began taking more of an active role in managing their energy usage. Their commitment to conserve and make decarbonizing lifestyle changes moved beyond household energy budget considerations and began to take on a patriotic fervor. Public utility commissions implemented dynamic pricing of electric energy nationwide.

In an effort to ease spring flooding in the Midwest and provide water to the thirsty West, several projects were completed to move water from the eastern drainage of the Rockies to the western states. Multiple water conservation policies including one of the more controversial measures, the banning of garbage disposals, were adopted in drought prone areas.

Nevertheless, water continued to be a vexing problem: too much in some regions and too little in others. Worldwide, the direct use of fossil fuels to produce freshwater in flash distillation plants ceased. Only electrified reverse osmosis plants continued in service and their numbers increased.

Efforts to preserve and restore wildlands ramped up. The hallmark Rainforest Protection Initiative by the UN gradually halted and then reversed decades of losses of tropical rainforests. The effort was successful only because of the worldwide boycott of Brazilian beef and punitive trade actions against any countries that didn't join the boycott. Sadly, the efforts to preserve wildlife habitats came too late to save the last wild elephant herd in Africa. The great beasts fell victim to higher temperatures, habitat loss, poaching and disease. Elephants survived only in zoos and books; breeding programs to rebuild the herds are on-going.

In the two decades since the carbon tax was implemented, per capita meat consumption in the U.S. decreased by 19% as a result of higher meat prices and the market success of plant-based meat substitutes. Over that same period, obesity rates in the U.S. dropped by 7%, but still remained the highest in the world. The nation achieved an average weight loss by adults of eight pounds (primarily as a result of "sin taxes" on high fructose corn syrup).

Waste materials flowing into landfills were cut by 80% as "cradle-to-grave" manufacturing was implemented. Virtually everything was recycled, reprocessed, remanufactured, or consumed in biomass-burning generating stations.

Under a UN initiative, each nation established its own population targets and the methods to achieve them. The Catholic Church even approved proactive family planning and contraception. Countries that met their family goals were rewarded with food aid, medical assistance and economic development, all paid for with proceeds from the carbon tax.

Weather-driven disasters continued to plague the planet. The frequency of extreme weather events became so regular that the U.S. established a National Emergency Response Corps, (NERC), which could best be characterized as a permanent National Guard force equipped and trained to respond to natural disasters including floods, earthquakes, forest fires, tornadoes and hurricanes. Loosely modeled after the Pan-European Fire Fighting force, the NERC grew to a force of 12,000 and acquired a fleet of fifty-eight fixed wing fire-fighting aircraft and 220 helicopters. Next year the Corps will celebrate its twenty-fifth anniversary. In twenty-three of those twenty-five years, the Corps responded to emergencies in all fifty states. Between tours on the front lines of disasters, the mostly young NERC forces undergo disaster training, aggressively conduct prescribed burns (which, before NERC, fire crews could not conduct because they were simply too exhausted from fighting seasonal fires) and enjoy generous vacation time. They don't receive hazardous-duty bonus pay for what they do because all their assignments are hazardous.

Perhaps the only good things to come out of the COVID-19 pandemic were the changes in the workplace and classrooms. Working from home a minimum of two days per week became standard practice for the majority of office workers. Currently about 17% of all office workers work from home 100% of the time.

Education also shifted gears after the pandemic of 2019-2022 receded from the headlines. Voluntary, universal pre-K for 3- and 4-year-olds was implemented nationwide. Beginning in the 4th grade, students started receiving 20% of their instruction through computer interfaces and 80% in smaller in-person classes. Junior high school students received 40% of their instruction via computer and 60% in person while high school students were 60%/40%. Most colleges adopted an 80% online with 20% in-person approach. After a shake-out period, educational achievement scores began a slow upward trend that has continued to this day. Changes in work and school models resulted in much less traffic on the roads. This made commuting for those who were unable to work from home a lot faster.

Energy consumption and the emissions associated with online shopping also dropped dramatically. Overnight deliveries became a thing of the past. They still exist (in theory) and are known as "Right Now" deliveries, but they carry a hefty surcharge of $40 per item. All non-perishable, online purchases are now coordinated and consolidated such that any given household receives a maximum of two deliveries per week. A computer system controls purchase fulfillment to ensure all items ordered from all vendors are bundled for delivery at the same time. The industry changes decreased air transport, increased rail transport, reduced cardboard consumption (and disposal), dramatically reduced the "final mile" expenses and reduced emissions.

Conservation and the greater deployment of carbon-free resources (solar, wind, nuclear and hydro) slowed the rate of increase in the concentration of CO_2 in the atmosphere. Experts predict that GHGs will peak out at around 480 ppm in 2055 and then begin a slow but steady decline as more nuclear plants come online, the last fossil-fuel generating stations are shut down and CO_2 is scrubbed from the atmosphere by natural and man-made processes. Some scientists are still lobbying for geoengineering measures (specifically injecting sulfur aerosols or precursor gases into the upper atmosphere) to hasten the return to an equilibrium heat balance at a lower average temperature. Most climate scientists, however, argue that the success of GHG reduction programs has negated the need to employ geoengineering techniques that could have adverse environmental effects.

Scenario #2 vs. Scenario #6

As you read Scenario #6, did you catch yourself thinking, "Oh, that'll never happen!"? Every time you had that reaction, another brick got pulled out of the foundation of a potentially successful, credible response to global warming. If you had a lot of those responses, it's possible you may agree with me that we're not going to "beat" global warming. You may even agree that we need to do something dramatically different than what we have been doing to respond to the crisis.

So, which of these two scenarios, Scenario #2 (The U.S. and the World "Snooze") or Scenario #6 (The World Wakes Up), do you believe is more likely to unfold in the next few decades?

You know my answer and while I hope that my arguments have been clear enough and the reasoning sound enough to allow you to come to the same conclusion, albeit perhaps begrudgingly, I know that some of you are not yet on board with this "we-need-to-do-a-lot-more" bandwagon. And I also know that fictional narratives like the two scenarios I laid out are not going to be enough to bring many of you around to my conclusion.

In the final analysis, my words (and your acceptance or rejection of them) don't really matter. What does matter is what's happening on the ground, or, in the case of global warming, what's happening in the atmosphere. **What matters are the DATA, the real-world DATA.**

> *My words (or your acceptance or rejection of them) don't really matter. What does matter is what's happening in the atmosphere. What matters are the DATA.*

Educator and management consultant, Peter Drucker, became famous using data to help organizations perform better. One of his credos, **"What gets measured, gets managed"** is especially applicable to CGI. The exhibits on the next two pages list some key parameters that should be measured on a continuing basis to shed some light on how much progress we're making, or not making, to avert the worst consequences of global warming. It's the data in these tables (or tables like them) that will matter over the decades to come. Every rosy report about how many new solar panels have been deployed or wind turbines made operational won't matter if the atmospheric concentration of CO_2 continues to climb and if the trend lines for some of these key parameters continue moving in the wrong direction. That's what we need to focus on. In the end, that's all that really matters. If you're not convinced by my arguments, at least be open to watching the data unfold. Let the DATA be your truth-teller.

Exhibit 19.2 is a set of general, worldwide measures for tracking global warming. Exhibit 19.3 is a set of parameters focused on the U.S. These lists are by no means exhaustive. For this first edition of Blue Oasis No More, I wanted to keep each list to just one page. I'll gladly expand these lists, substitute more critical metrics and fill in the numbers in a future edition.

Exhibit 19.2 Worldwide Parameters of Interest

Parameter	2019	2030	2040	2050
Worldwide Annual CO_2 Emissions (10^9 Tons)				
Atmospheric Concentration of CO_2 (ppm)				
Annual Worldwide Coal Production (10^9 Tons)				
Ocean Acidity (pH)				
Amazon Rainforest Area Lost (10^6 Square Miles)				
Worldwide CO_2 Extracted and Stored (10^6 Tons)				
Avg. Earth Temperature (°C)				
Worldwide Carbon Tax ($/Ton)				
Number of Nations Reducing GHG Emissions By At Least 5% Relative to 5 Years Earlier (#)				
Worldwide "Weather" Impacts (deaths, property damage, etc.) Scale 1 – 10 (1=Low, 10=High) (Impacts in 2019 = 1)				

Exhibit 19.3 U.S. Measures of Interest

Parameter	2019	2030	2040	2050
U.S. Total Electrical Energy Consumed (All Sources Including Self-Generation, 10^9 kWh, Billions of kWh)				
U.S. Annual CO_2 Emissions (10^9 Tons)				
Percentage of U.S. kWh Produced by Wind and Solar Resources				
Percentage of U.S. kWh Produced by Nuclear Resources				
Percentage of U.S. kWh Delivered from Battery Storage				
Annual U.S. Weather-Related Deaths (Heat, Cold, Tornadoes, Forest Fires,Hurricanes,etc.)				
Annual U.S. Per Capita Meat Consumption (pounds)				
Annual U.S. Consumption of Gasoline, Diesel (10^6 Gallons)				
EVs As A Percentage of all U.S. Cars/Light Trucks on the Road				
Total Annual U.S. Consumption of Natural Gas, (10^9 therms)				
Annual U.S. Natural Gas Consumption Excluding Electric Generation (10^9 therms)				
Annual U.S. Domestic Coal Consumption (10^6 Tons)				
U.S. Electric Grid Reliability %				

Scenarios and Scorecards

20 TOMORROW'S FORECAST: SCORCHING

"It's Too Darn Hot"

Kiss Me Kate
Song Lyrics by Cole Porter, 1948

In the Introduction, I noted that the simple question at the heart of this book, "Will we 'beat' global warming?" really had three fundamental or implied elements.

1. GHGs: Will we get to zero emissions and will we reduce the atmospheric concentration of GHGs?
2. Temperature: How hot will it get? And,
3. Life as we've known it: Will the increase in global temperatures truly impact the quality of life on planet Earth?

Thus far, I've only addressed the first question. I've made the case that **we will not reduce CO_2 emissions quickly enough and we will not drawdown the atmospheric concentration of GHGs sufficiently to stop global warming.**

Given this answer to the GHG question, the second question takes on greater importance: "How much will the global average temperature increase as a result of the added GHGs in the atmosphere?"

My experience in energy research, production, conversion and usage gives me a unique perspective to add something of value to the first question regarding CO_2 emissions from energy-related activities, but I'm decidedly less qualified to address the second question. Nevertheless, I'll take a crack at it using a simple and, I hope, understandable analysis of a complex question. My approach assumes a linear relationship between earth's temperature rise and the magnitude of the excess CO_2 being carried in the atmosphere over time. In other words, if x causes y, then 3x will cause 3y. Follow along with me.

In Chapter 3, I described how a higher concentration of CO_2 in the atmosphere traps and retains more of the sun's energy. Based on those principles, it makes sense that more energy will be trapped and retained, at least temporarily, inside the earth's atmosphere for higher concentrations of CO_2. The absorption of that extra energy will cause the rate of global warming to increase. This will continue until those higher concentrations of GHGs are brought down or until the earth heats up to a temperature high enough that the amount of energy radiated back into space equals the extra energy being absorbed by the earth.

In the thousands of years leading up to the industrial age, the atmospheric concentration of CO_2 corresponding to a stable earth temperature was roughly 280 ppm. In this analysis, I refer to this as the baseline CO_2 concentration. Values in excess of this equilibrium level are "excess CO_2 concentrations." The extra energy trapped inside our atmosphere is proportional to the magnitude of this excess CO_2 and the length of time the excess CO_2 has resided in the atmosphere. In mathematical terms, the time-integrated excess CO_2 concentration (excess CO_2 ppm-years) is proportional to the increase in global average temperature. Knowing the temperature increase the earth experienced in the period from 1900 to 2000, we can estimate the global temperature increase that will occur from 2000 to 2100 based on the forecasted time-integrated concentration of excess CO_2 over that period.

Climatologists have conservatively estimated that the global average temperature increased by approximately 0.8° C (1.4° F) over the 100-year period from 1900 to 2000 as a result of excess CO_2 in the atmosphere over that century. Most of that increase occurred in the last two or three decades of the 1900s when the excess CO_2 was at its highest.

Exhibit 20.1 shows the atmospheric concentration of CO_2 over the period from 1900 through 2100. The data from 1900 through 2020 are recorded numbers. They're real. The values from 2020 through 2100 are estimates using different assumptions about how successful we'll be at decreasing emissions.

Exhibit 20.1
CO_2 Concentration With Assumptions
About Future Emissions

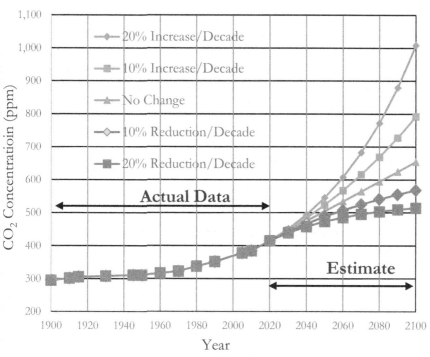

In the period from 2010 to 2020, the concentration of CO_2 in the atmosphere was increasing at a rate of approximately 3 ppm/yr. I've explained why I don't believe cutting CO_2 emissions to zero in thirty years is attainable. However, I also believe that more and more renewables will be deployed in the U.S. and around the world every year and this, along with all of the other CO_2 emissions reduction strategies (e.g., conservation, energy conversion efficiency improvements, EVs, etc.) will result in modest reductions in the rate at which CO_2 builds up in the atmosphere.

The projected concentrations of CO_2 I'm showing in Exhibit 20.1 over the period from 2020 through 2100 reflect five different assumptions regarding future CO_2 emissions. The middle curve assumes we continue pumping CO_2 into the atmosphere at a rate resulting in a net increase of 3 ppm per year. In other words, the net effect of everything we do to reduce CO_2 emissions is offset by other factors causing increases. (I'll explain shortly how this could happen.) The two lower curves reflect CO_2 emissions reductions of 10% and 20% per decade. These two curves reflect what might develop if the world pursues aggressive programs to transition to carbon-free energy resources. Note that the CO_2 concentration curves for both the 10%/decade reduction and the 20%/decade reduction scenarios, continue to go up but at a slightly slower rate.

The other two curves in this exhibit show increases in the rate of CO_2 buildup in the atmosphere of 10%/decade and 20%/decade. You might think it's misguided to show potential CO_2 emissions increases in future decades just as the world is waking up to the reality of global warming. These curves reflect the real possibility that we may continue to make great strides in reducing our consumption of fossil fuels and simultaneously see an acceleration in the increase of atmospheric CO_2. How could that be?

The measured concentration of CO_2 in the atmosphere is a function of mechanisms that both add to and subtract from that concentration. If the removal mechanisms become less effective or the adding mechanisms

> **The rate of global warming could accelerate (_not decelerate_) in coming years in spite of efforts to reduce GHG emissions.**

increase, the concentration of CO_2 in the atmosphere could go up in spite of our best efforts to reduce human-generated emissions. Some of the factors that could cause increases in CO_2 or increased warming are:

1. There may be limits to the quantity of CO_2 that the oceans can absorb at the air-water interface.
2. The continuing conversion of rainforests to agricultural uses and vast forest fires may reduce the net amount of vegetation and the effectiveness of that vegetation to pull CO_2 out of the atmosphere and store it.
3. Higher average temperatures and an increasing frequency of excessive heat events may cause once-healthy trees to drop their leaves prematurely or just die.
4. The warming and acidification of the oceans could kill massive quantities of marine life, causing the release of GHGs in the decaying process.

5. Tundra regions around the world are expected to warm sufficiently to allow previously frozen organic matter to decay at an accelerated rate releasing large quantities of methane gas.
6. Dust and dirt stored in layers of ice will accumulate on the surface of glaciers as they melt. As the surface of the ice gets dirtier and darker, those surfaces will reflect less energy back into space.
7. As sea ice and land ice melt, the dark ocean and soils that were previously covered by reflective sheets of ice are now exposed and will absorb more of the energy in the incoming sunlight.
8. Increasing populations, especially in developing countries, will require more goods and services that will drive up emissions.

If some sources of emissions increase and natural mechanisms to remove CO_2 become less effective, the buildup of GHGs and global warming may not abate. We'll probably know which way the GHG concentration and global warming curves are trending within a decade, but as of now, we have to at least acknowledge the possibility that things could get worse despite significant efforts to decrease emissions.

I'm hopeful that the net effect of changes in the CO_2 removal mechanisms and CO_2 adding mechanisms will be negative, resulting in a decrease in the rate at which CO_2 is building up in the atmosphere. (If you've read Appendix B, you know that deep down, I'm an optimist).

The Estimate

For the purposes of this book, I assume that we, humankind, will be modestly successful and we'll achieve the equivalent of a 10% per decade net reduction in the rate at which CO_2 increases in the atmosphere. Achieving even this modest outcome will be no slam-dunk in a world where the population is increasing, where efforts continue to raise the standard of living of billions of people around the globe and where the starting point for making these emissions reductions is an emissions rate that had previously been increasing. Nevertheless, a 10% per decade reduction in the rate of increase of CO_2 in the atmosphere would decrease the 2019 rate of increase of 3 ppm/year down to 2.7 ppm/year by 2030, 2.43 ppm/year by 2040 and 2.19 ppm/year by 2050.

Using these estimates for the rate of increase of CO_2 in the atmosphere, Exhibit 20.2 shows the actual and projected atmospheric CO_2 concentration through the year 2100. [Note: Exhibit 20.2 is just an enlarged and simplified version of the estimate in Exhibit 20.1 showing a 10%/decade decrease in the rate of the increase in CO_2. The "cross-hatching" has been added to highlight the excess CO2 in the atmosphere over time.] The **area under the CO_2 concentration curve and above the 280-ppm concentration level from 1900 to 2000** (i.e., the cross-hatched area to the left of year 2000) **provides a quantitative estimate of the "excess-CO_2-ppm-years" that gave rise to the 0.8° C temperature increase that occurred over that century.** By calculating the comparable area for the period 2000 through 2100 (i.e., the cross-hatched area to the right of year 2000), it's possible to make an estimate of the temperature increase the world will experience in this next century based on the projected CO_2 concentration increases.

Exhibit 20.2

Cross-Hatched Area Proportional to Global Temperature Increase
(Assuming Atmospheric CO_2 Buildup Decreases 10% Per Decade)

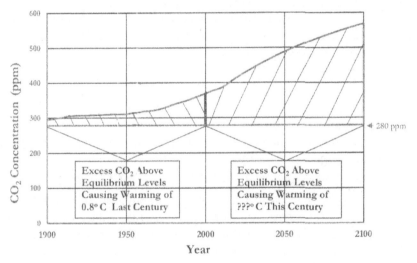

Those calculations reveal that the "excess-CO_2-ppm-years" for the period 2000 through 2100 is about four times larger than the comparable area for the period from 1900 through 2000. If the global average temperature increase that the earth experienced in the 20th century was 0.8° C (1.4° F), this "analysis" suggests the earth is likely to experience four times that increase, an additional 3.2° C (5.8° F), in this century. (And remember, this **assumes** we (the world) will **reduce**

net CO_2 emissions by 10% per decade for the remainder of the century). The combined global average temperature increase over this two-century period would be 4° C (7.2° F).

OK. This is not an overly sophisticated analysis, but it actually lines up reasonably well with predictions made by the experts for these estimated levels of atmospheric GHG concentrations. I hope it's at least both logical and understandable. I've included it not so much for the calculated result, but mostly to illustrates the two crucial factors affecting global warming and excess energy retention inside our atmospheric envelope. The two factors are the magnitude of the excess atmospheric concentration of CO_2 (above historical, "equilibrium" levels) and the length of time that excess CO_2 has persisted in the atmosphere. Using these two factors, it's easy to see that the area under the excess CO_2 curve from the year 2000 through 2100 (the cross-hatched area to the right of the year 2000) is much greater than the cross-hatched area under the curve from 1900 through 2000. The bottom line: future temperature increases are likely to be much larger than what we've experienced thus far - unless we're able to get our GHG emissions down dramatically and see a reduction (by man-made and natural processes) in the excess CO_2 in the atmosphere. If we don't (and I've argued over the last 200 pages that we won't) our grandchildren will be coping with a much warmer world and all the mayhem that will bring.

This outcome, by the way, is not "negotiable." These temperature increases will be a function of invariant physical laws. No amount of "wishing it weren't so" or politically inspired rhetoric to the contrary will make the outcome any different. Climate experts have predicted dire consequences if the earth warms 1.5° or 2.0° C above what it had been for thousands of years. They've made their predictions public in a desperate attempt to get all of us to act collectively and quickly to curb GHG emissions.

It's not working.

Given the weather impacts we've already experienced with only a 1° C increase in the average temperature of the earth through 2020, you don't have to be a climatologist to imagine that life on an earth that's at or near 4° C (7.2° F) hotter than it's been for the last 100,000 years is going to be quite a bit less pleasant than what most of us may have "enjoyed" in the fossil-fuel era. That's the abbreviated "answer" to the third element of the question: "Will we 'beat' global warming?"

The combined answer to all three elements of the question is: **we will not reduce CO_2 emissions quickly enough to meaningfully slow global warming and as a consequence the global average temperature will continue to rise year after year for the foreseeable future dramatically altering life on Earth.** We will not "beat" global warming and future generations will curse our inaction.

A recent national newscast warned listeners to prepare for a "summer of extremes." The news anchor might just as well have warned his listeners to prepare for a "century of extremes."

In a radio interview on NPR, climate action advocate Bradley Udall characterized these coming changes in the starkest of terms. His description may provide a **transformative** glimpse into the future.

Mr. Udall urged us to stop thinking about the record temperatures in the summer of 2021 as the "hottest on record." Instead, he urged us to think about these record-breaking temperatures as **the coolest of the next century**. Think about it. If that doesn't stop you in your tracks, nothing in this book will.

The changes a warmer climate will unleash on us will **NOT** bring an end to life on earth, but they will cause enormous and, in time, truly painful changes. Our once-thriving, rocky, little speck in the universe will be our **Blue Oasis No More.**

21 FIFTEEN "MYTHS"

"I want the truth." Lt. JG Kaffee (Tom Cruise)
"You can't handle the truth." Col. Jessup (Jack Nicholson)

A Few Good Men, 1992

There are those who don't believe global warming is real. They're wrong and the fact that you're reading this book probably means you feel the same way.

However, there are also a number of "myths" that are widely believed by those of us who are concerned about global warming and desperately want us (the collective "us") to do something about it. Perhaps you believed some or all of these "myths" before reading this book. If you now recognize them as not being true, then I've done at least part of what I set out to do. Here they are:

1. Energy from a grid powered by renewable resources will be cheaper than energy from a grid powered by fossil fuels. (Chapters 13 and 15)

2. EVs are zero emissions vehicles. (Chapter 11)

3. Storing carbon-free energy generated during the day for use at night is the best strategy to reduce CO_2 emissions. (Chapter 9)

4. Once we stop burning fossil fuels, we're home free. (Chapters 12 and 13)

5. The window (to address global warming) is closing. (Chapter 4)

6. Solar panels, wind turbines, EVs and transmission lines -- no problem -- we've got this.
(Chapters 4, 8, 9, 11, 12 and 13)

7. Once we finally get to zero CO_2 emissions, global warming will be solved.
(Chapters 3 and 4)

8. Once we make the transition to renewables and energy storage, life can go on pretty much as it has for the last 50 years.
(Chapters 17 and 20)

9. We can make the transition to a carbon-free energy future in an equitable way for all the world's people.
(Chapters 13, 15, 17 and, 20)

10. We are the first generation to feel the effects of global warming and the last one that can do anything about it.
(Chapter 4)

11. The world will finally swing into action after major storms ravage Miami, Cancun, Shanghai, Mumbi, Rotterdam and Dacca . . . all in one year.
(Chapters 14, 15 and 19)

12. Technology will fix this.
(Chapters 7, 8, 9, 11, 12 and 16)

13. Nuclear power can fix this.
(Chapter 7)

14. Young people will fix this.
(Chapter 17)

15. If we do x, y, and z, we can "beat" global warming.
(All Chapters)

22 MY CLOSING ARGUMENT

"We have not inherited the earth from our forefathers,
We're borrowing it from our children."

"If you borrow something you don't have the
capability of paying back, [you're]
actually stealing."

The first quote is one of several variations characterized
as an Amish saying, a Native American saying
or credited to a host of individuals.

The second observation is by Environmentalist
David Brower, 1989

Are we going to "beat" global warming?   Recall that this
question is shorthand for the more precise (but wordier) version:

> *Will we reduce our greenhouse gas emissions and*
> *their concentration in our atmosphere*
> *quickly enough to avoid*
> *major, recurrent, weather- and climate-related*
> *economic losses, human losses*
> *and ecological damage*
> *so that life, as we've known it,*
> *can flourish in the future?*

I started this book by giving my answer to this question,
without any supporting arguments. My answer then, and still is:

> ## No, we will NOT.

Having now presented the supporting arguments that have led me to this unhappy conclusion, **what do you think?** Will we, or won't we?

Before you answer, let me briefly recap the primary elements of my argument.

The CO_2 Spigot Is Wide Open

Worldwide, the burning of fossil fuels (coal, oil, and natural gas) is releasing an estimated 40 to 50 billion tons of CO_2 into the atmosphere every year (Chapter 4). Before the start of the industrial age (and for thousands of years prior to that) the concentration of CO_2 in the atmosphere was fairly stable in the range of 280 ppm. Today it's 420 ppm, 50% higher, and increasing at the rate of roughly 3 ppm per year. It could reach 500 ppm by 2050.

Once released, CO_2 that's not absorbed in the ocean remains in the atmosphere for hundreds of years. As the atmospheric concentrations of CO_2 and other greenhouse gases (GHGs) increase, the fraction of solar energy impinging on our atmosphere that's trapped and retained by the earth increases. As more energy is retained by the earth, the temperature of the air, water and soil increase (Chapter 3) and will continue to increase until a new hotter equilibrium temperature is reached or the atmospheric GHG concentration is reduced.

So, we have two problems. The first is that we're making the problem worse every day by continuing to release GHGs into the atmosphere. By adding to the GHGs that are already there, **the rate at which the earth is heating up is increasing.** We need to stop dumping CO_2 into the atmosphere.

The second problem is the excess CO_2 we've already dumped into the atmosphere. If we were somehow able to reduce our global warming emissions to zero (which I've argued we will not) we'd still have a problem. The excess GHGs already in the atmosphere will continue trapping more of the sun's energy and the earth will continue to heat up until a new, higher equilibrium temperature is reached. Eventually we need to bring down the atmospheric GHG concentrations using a combination of natural and man-made processes, which we also won't be able to do.

If a new, higher equilibrium temperature materialized over hundreds of thousands or millions of years, perhaps global warming wouldn't be a problem. But the global warming caused by human activity is occurring over a period of decades and will have catastrophic consequences for all forms of life on the planet.

For the past seventy years, researchers have been actively seeking the holy grail of energy: a clean, cheap, inexhaustible energy technology. Initially, the motivation was the fear that fossil-fuel resources would run out. In the 1980s and '90s, however, scientists began recognizing global warming was a greater, more imminent threat.

As we enter the third decade of the new millennium, our knowledge of the science of global warming and the looming consequences of climate change have expanded greatly, but the tools we have to reconfigure the world's energy economy have not.

Nuclear fusion power is now, **and will always be**, one lifetime in the future (Chapter 16). Geothermal energy

> *We're going into the third decade of the new millennium with painfully few arrows in our quiver (solar panels and wind turbines) to attack the power generation dimension of global warming.*

is potentially viable, but very site-specific and has significant environmental issues. The production of ethanol and other biofuels is already limited and is likely to shrink as worldwide food demand increases. Garbage and biomass incineration will be more widely utilized, but these generation technologies present a myriad of environmental, safety and public acceptance concerns. Hydroelectric power is proven, cost effective, dependable, and flexible, but most of the best hydroelectric sites (that don't have significant environmental issues) have already been developed (Chapter 8). Ocean wave power is too expensive and too vulnerable to storms and everything else the unforgiving ocean environment can serve up.

Nuclear (fission) reactors are carbon-free, proven, cost effective, safe, utilize a virtually inexhaustible fuel supply, and are a perfect complement to solar power (Chapters 7 and 10) to serve a typical mix of customer energy demands. Nevertheless, the general public remains wary of nuclear power and political leaders lack the knowledge or courage (or both) to make it part of the solution. Nuclear power will remain the energy option of last resort in most countries until after 2050 or 2075, i.e., well past the point when the worst consequences of our catastrophic global imbalance (CGI) begin taking an irreversible toll.

> *Nuclear power will remain the energy option of last resort in most countries until after 2050 or 2075.*

Without a carbon-free, storable fuel, the only strategy available to "beat" global warming is to electrify virtually everything we do with carbon-free energy resources (Chapters 4, 6, 8, 9, 10, 11 and 12). Without the nuclear power option, the two primary energy technologies available to accomplish this worldwide energy decarbonization miracle are wind turbines and solar panels.

These renewable energy sources are carbon-free, proven, inexhaustible and, when producing, can generate cost-competitive energy. They are also accepted and supported by a substantial portion of the public (Chapter 8). These technologies are now, and will continue to be, in the forefront of worldwide efforts to electrify everything. In the years ahead, utility grids around the world will add enormous quantities of both solar and wind power.

Unfortunately, both of these renewable technologies have the same pesky drawback: they're both intermittent resources. Solar panels have a predictable unavailability at night and a variable unavailability because of clouds and other weather factors.

> *Unfortunately, solar panels and wind turbines have one pesky drawback: they're both intermittent resources.*

The intermittent nature of wind and solar resources means we'll have to make massive investments in the electrical grid of the future. We will have to add enormous quantities of bulk energy storage, the additional carbon-free resources to recharge that storage plus the high-voltage transmission infrastructure to move energy long distances.

Without enormous investments in storage and transmission capacity, the only options available to utilities to serve their customers when the intermittent technologies are unavailable is to burn fossil fuels or use brownouts and blackouts as an operational strategy to match supply and demand. Using the fossil fuel option on an "as needed" basis is inefficient, hard on the equipment, emits more CO_2 per kWh, and continues the human-driven assault on the atmosphere. The brownout/blackout option would be unacceptable in modern, developed economies.

In all likelihood, most utility systems will **not** deploy sufficient bulk energy storage capacity to become **totally independent** of fossil generation. But this means CO_2 emissions will not go away and utilities will carry the added expense of maintaining renewable resources and energy storage capacity alongside their original fossil-fuel resources. If those fossil-fuel-fired resources are still available to be called into service as needed, there's the added environmental risk that it will become all too easy for regulators, under pressure from the general public, to fall back on those cheaper (CO_2-producing) resources when the alternative is an expensive combination of energy storage and renewable generation.

If utilities wanted to completely decarbonize the fossil-fuel portion of the existing U.S. electrical grid, they would have to deploy the equivalent of 4.7 billion residential-size solar panels or 287,000 medium-size wind turbines (Chapter 10), together with massive quantities of bulk energy storage and thousands of miles of new transmission lines. Some estimates of decarbonizing the existing electrical grid are in the range of 10 to 15 trillion dollars with an additional and roughly equivalent investment for energy storage plus the added transmission infrastructure. All this to decarbonize just the existing U.S. electrical grid, which accounts for less than a third of the fossil fuels consumed annually in the U.S.

Beyond the Current Electrical Grid

In addition to the existing electrical grid, all of the non-electric-generating consumption of fossil fuels in the transportation, residential, commercial, and industrial sectors will have to be replaced by carbon-free energy sources (Chapters 11 and 12).

The transportation sector presents unique challenges. The average life of cars on the road in the U.S. today is between ten and fifteen years. That means that most of the new gasoline-powered vehicle rolling off the assembly lines today are likely to still be on the road in 2035. With annual car and light truck sales in the U.S. of 17 million vehicles in 2019 and the 77 million vehicle sales around the world it should be clear that gas stations will not be an endangered species for several decades. Annual EV sales will increase rapidly, but EVs will probably account for no more than 25% of all personal transportation vehicles on U.S. highways by 2035. In other words, the transition away from fossil-fuel-powered transportation will take a long time.

The second problem related to eliminating CO_2 emissions in the transportation sector is that EVs are not truly zero-emissions vehicles. They are, more accurately, **zero tailpipe-emissions vehicles**. The energy to recharge EV batteries still has to come from somewhere. The vast majority of the energy flowing into EV batteries has and will continue to come from the electrical grid.

> *EVs charged off the grid today "emit" (are responsible for) almost 80% as much CO_2 as today's high-mpg gasoline-powered vehicles.*

Only after the grid is 100% carbon-free will it be accurate to characterize EVs be zero-emissions vehicles. Until then, EVs recharged off the existing power grid (i.e., burning natural gas as the marginal fuel to generate electricity), emit only about 20% less CO_2 than today's high-mpg vehicles (i.e., cars that get forty-five miles or more per gallon). As gasoline-powered vehicle fuel efficiency improves the marginal emissions benefit provided by EVs over gasoline-powered vehicles will actually shrink (until the grid is almost entirely "green.")

All the electric transportation options that the country will employ to reduce CO_2 emissions (E-cars, E-pickups, E-light duty trucks, new mass transit, new electrified freight trains, etc.) represent a **new electrical load** that will require the construction of even more carbon-free electric generating capacity, additional energy storage and carbon-free generating capacity dedicated to charging that storage. Serving this new electrical load will require the addition of roughly as many new renewable resources as would be needed to decarbonize the existing electrical grid. To serve this new electrical load, 3.8 billion residential-style solar panels **or** 236,000 medium-size wind turbines, plus massive quantities of energy storage and thousands of miles of new transmission lines (Chapter 10) would be required.

Process Heat

The final portion of fossil-fuel consumption that needs to be replaced by clean energy resources is the energy derived from burning fossil fuels directly in our homes, businesses and industries (e.g., in hot water heaters, ovens, kilns, boilers, furnaces, etc.). The carbon-free resources necessary to replace this component of fossil-fuel consumption in the U.S. is roughly two and a half times the amount of capacity needed to decarbonize our existing electrical grid. And, crucially, replacing this fossil-fuel-burning equipment with electric alternatives can't begin **until** our expanded national electrical grid is almost completely carbon-free. This is because most of this fossil-fuel-burning equipment is at least twice as efficient as their electric alternatives. In other words, replacing fossil-fuel-burning equipment in our homes, business, and industries with an electric alternative **before** the grid is 100% green could **double CO$_2$ emissions** attributable to those functions. (Chapter 12).

The Big Picture

The scale of the challenge to decarbonize our existing electrical grid, our transportation sector and our residential, commercial, and industrial activities is best appreciated by looking at some simple numbers. Eliminating fossil fuels from just the U.S. energy economy would require deploying:

Over 20 billion (residential size) solar panels[1],
or
1.3 million medium-size wind turbines,
or
1,180 large nuclear power plants[2]

---

1. That's roughly 62 solar panels for every man, woman, and child in America. (A typical household rooftop solar installation currently consists of 20 to 40 solar panels).
2. The numbers for the nuclear option are shown for comparison purposes only, because, as I've argued, the nuclear option, for now, is pretty much off the table as far as the public is concerned. [Note: These numbers are provided to illustrate the magnitude of the challenge. I'm not suggesting that we could or should deploy specific renewable resources in these numbers.]

If per capita energy consumption remains the same as it is today, all these numbers would increase by at least 20% to 30% due to population growth through 2050 and the need to replace carbon-free resources that are currently in operation as they reach their end of life or are taken out of service for any reason. If the transition away from fossil fuels is to be accomplished using only intermittent technologies, i.e., wind and solar resources, the transition will also require a massive investment in energy storage and high voltage transmission lines.

We Will Do EVERYTHING and It Still Won't Be Enough

In reality, efforts to slow global warming (by reducing CO_2 emissions) will (or at least, should) employ every strategy and technology we can muster: a carbon tax, hydroelectric, solar panels, battery storage, wind turbines, conservation, nuclear power, pumped hydroelectric storage, geothermal, energy efficiency, dynamic energy pricing, EVs, mass transit, high voltage transmission, biomass incineration, urban planning, heavy rail electrification, ethanol, lifestyle changes, reduced meat consumption, tree planting, soil management, wetlands restoration, forest protection, carbon capture, green hydrogen, ammonia and perhaps another dozen I've either forgotten or don't know about. Different countries will employ different combinations of technologies and strategies to lower their carbon footprints. Some technologies, like nuclear power, will be underutilized in most countries (in the near term) due to a lack of public support and unwarranted expectations that other technologies (especially wind and solar) will suffice to solve the problem.

> *Efforts to slow global warming by curtailing CO_2 emissions will (or at least, should) employ every strategy and every technology we can muster.*

We'll use all these tools, but in the end, we won't be able to reduce atmospheric GHGs enough, and not quickly enough, to avoid the worst consequences of global warming. It's entirely possible that we'll blow past the 1.5° C temperature increase limit climate scientists have been warning us about as soon as the late 2030s and maybe even double that increase by sometime in the 2070s.

A recent McKinsey and Company report, "The Net-Zero Transition – What it Would Cost and What it Could Bring" estimated the worldwide expenditure to limit the global temperature increase to 1.5° C would be $9.2 trillion . . . **per year** through 2050!

More Than Technology

While technology is important and is the element that most people focus on, our ability to adequately respond to our catastrophic global imbalance is not exclusively or even primarily a function of technology. Economics, politics, education, culture, psychology, emotion, **and** technology all have a

> *Economics, politics, education, psychology, culture, emotion, and technology all have a bearing on the world's response to global warming.*

bearing on the world's response to global warming, and we're not "winning" in even a single one of these dimensions of the problem.

"Enemies" in High Places (Including Us)

All the strategies to drive down CO_2 emissions have faced (and will continue to face) powerful headwinds. Vested interests will resist changes that diminish the power or profitability of their enterprises (Chapter 14). And just to be clear, those vested interests are not limited to corporate executives wearing black hats. They include shareholders who want their investments to excel, anyone unwilling to prematurely scrap fossil-fuel-burning durable equipment and all of us who want to continue the energy-intensive lifestyles we've gotten used to over most of our lives. They, i.e., **all of us,** will do what we can, sometimes unconsciously, to perpetuate the status quo. As energy costs rise, vital support for making near-term sacrifices to achieve far-in-the-future rewards will fade. Other urgent needs will cause governments and individuals to lose focus and redirect resources away from long-term decarbonization efforts (Chapter 14) to more immediate needs.

Combating climate change is also going to require the entire world to focus intensely on a single target for decade after decade, something that's never happened before (Chapter 14). Worldwide cooperation and progress to "beat" global warming will also require something akin to an international decarbonization authority or treaty agreements that are so strong that individual nations effectively lose their ability to "opt-out" -- also something that's never happened before.

One Step Forward, Two Steps Back

The two most **encouraging** developments in combating global warming in the U.S. are: first, the decrease in the use of coal to generate electricity in the U.S. and second, the deployment of truly significant quantities of renewable energy resources. (However, it was the availability of cheap natural gas from fracking, not global warming concerns, that caused the accelerated decline of coal in the U.S. Additionally, the expansion of renewables shouldn't be over-hyped because intermittent renewable resources without adequate energy storage to make the grid "green" won't get the job done).

The two most **discouraging** developments related to global warming countermeasures are: first, the failure of the international community to enter into any form of a **binding commitment** to do what needs to be done including, most importantly, an international carbon tax (or Planet Survival Investment charge, a PSI), and second, the failure to move dramatically forward with advanced nuclear reactor technologies.

This is a multi-dimensional problem that we are not on track to solve. Virtually none of the crucial things that need to happen to "beat" global warming are happening and those that are, are happening too slowly to be effective.

> *This is a multi-dimensional problem that we are not on track to solve. Virtually none of the crucial things that need to happen to "beat" global warming are happening and those that are, are happening too slowly to be effective.*

Forget about asteroids crashing into the earth. We're on a collision course with ourselves and we're not making the mid-course corrections necessary to avert a disaster. We're on a path that may force us to try unproven and potentially harmful geoengineering approaches such as injecting sulfur aerosols into the stratosphere, a strategy that at present seems **unthinkable**. But even attempting such an extreme and as yet unproven countermeasure would require years of research and an international framework that doesn't now exist.

The best we can hope for from this point forward is that everything we do (e.g., deploying enormous quantities of renewable electric generation and all the rest) will slow the buildup of CO_2 in the atmosphere and buy us some time to allow the U.S. and the world to embrace a Planet Survival Investment (PSI) charge, aka a carbon tax, and advanced fission reactors. If we do those two things, we might be able to delay the onset of the worst consequences of global warming, but those "worst consequences" are coming. We're not going to "beat" global warming.

Twenty-Seven "Takeaways"

All authors hope that their writings have an impact on their readers, that they'll learn something new or at least gain a new perspective. Here are the points I hope all of you take away from this book. Some are obvious, some less so:

1.   Global warming is real. It's governed by the laws of physics and chemistry. Human activities adding greenhouse gases to the atmosphere are the biggest contributor to the observed warming.

2.   "Beating" global warming will require decarbonization of the world's energy economy. Virtually every energy-consuming activity will have to be electrified and powered by carbon-free resources or chemical reactions that don't increase GHG concentrations in the atmosphere.

3.   Energy in a decarbonized energy economy will be three to four times more expensive (in real terms) than it is today.

4.   Unless wealthy countries step in to provide assistance, decarbonizing the world economy will deprive at least half the world's population of access to the same kinds of energy resources that made the wealthy countries wealthy.

5.   Solar panels and wind turbines are the two most cost-competitive carbon-free renewable energy resources. They enjoy wide-spread public acceptance and will be installed in massive numbers around the world.

6.   The "Achilles heel" of solar and wind power generating resources is that they are intermittent. Developing a reliable, carbon-free electrical grid based on wind and solar resources will require a massive investment in bulk energy storage capacity **and** the additional renewables to produce the energy to flow into that storage.

7. Decarbonizing the world economy, however, will require a lot more than putting solar panels on every rooftop and an EV in every garage.

8. The energy and emissions benefits of EVs are being oversold. EVs charged off the electrical grid are not zero-emissions vehicles so long as **any** of the energy being supplied to the grid is coming from fossil fuels (coal or natural gas). The most popular EV in the U.S. recharged from today's electrical grid "emits" only about 20% less CO_2 (produced at the power plants) than a gasoline-powered car getting 45 mpg. Today's EVs are, in effect, high mileage vehicles powered not by petroleum, but by coal or natural gas. This characterization applies even if the EV owner has a solar panel installation and/or stationary battery system connected to the grid.

9. Burning natural gas by residential, commercial, and industrial customers is at least twice as energy efficient and emits about half as much CO_2 as processes that use electricity produced from fossil fuels. Converting process-heat applications to electricity before the electrical grid is completely carbon-free will **increase** CO_2 emission.

10. Technological fixes are not going to fundamentally alter the global warming trajectory. Even if breakthroughs emerge from the laboratory in the near future, it'll take time to refine the research, perfect the concepts, build prototypes, obtain operating data, refine designs, develop project plans, obtain financial backing, scale up production and deploy enough of the new technology to make a difference. And all the while, CO_2 will be building up in the atmosphere.

11. In the years to come, the media will be filled with stories about EV sales, ribbon cuttings at new renewable energy projects, the latest breakthroughs in fusion research, etc. Any progress to reduce CO_2 emissions will be beneficial, but what really matters is the concentration of CO_2 in the atmosphere and the global average temperature. Those are the two numbers to pay attention to.

12. Geoengineering concepts to shield the earth from a portion of the sun's rays are unproven at best, potentially harmful at worst, highly controversial and could be costly. They also do nothing to mitigate the damage to the oceans by CO_2.

13. **Any energy,** whether **from fossil fuels or renewable resources,** used to recharge an energy storage medium at a time when any fossil-fuel-generated energy is flowing into the grid to satisfy the immediate electrical demand will **increase CO_2** emissions compared to what they would have been if the storage system didn't exist.

14. Time is not on our side. The concentration of CO_2 has been rising at a rate of approximately 3 ppm per year. Monumentally successful efforts to curtail CO_2 emissions from human activities might possibly bring this number down to a net increase of 1.5 or 2 ppm per year by 2050, but the concentration of heat-trapping gases in the atmosphere will continue to go up for the foreseeable future.

15. The cumulative effect of the increasing concentration of greenhouse gases in the atmosphere will be a global temperature increase in the 21st century that could be four times as large as what was experienced in the prior 100 years. By the end of the 2030s we're likely to exceed the 1.5° C temperature rise limit climatologist have been warning us about. By the 2070s the global average temperature rise could be twice that amount -- even if we do much more to reduce CO_2 emissions than we have been doing up until now.

16. The degree of international cooperation necessary to "beat" global warming has never even been attempted, much less achieved. There's no basis for believing that this "first-of-a-kind" effort at international cooperation and trust will proceed smoothly.

17. The world's current population already exceeds the sustainable "carrying capacity" of the earth. The world's population is projected to increase by at least 25% by 2100. A growing population and improvements in the standard of living for

billions of people will drive up total CO_2 emissions even as decarbonization efforts attempt to reduce them.

18. As global temperatures increase, world food production will be adversely affected. Shifting weather patterns will expand areas of both drought and flooding. Producing adequate food and freshwater supplies to the world's population will absorb a larger and larger share of the world's attention and resources in the coming decades.

19. Powerful vested interests will attempt to derail emissions reduction efforts and will, from time to time, succeed. Establishing and maintaining majority support for decarbonization programs and policies that take money out of the wallets of ordinary people will be virtually impossible.

20. The earth is a dynamic system that was more or less "in balance" for many thousands of years before the industrial age. Human activities have disrupted that balance and set in motion a series of changes. Some of those changes and their consequences are "knowable;" many are not. None of the "unknowns" is likely to make the task of countering global warming any easier.

21. A sizeable portion of CO_2 emissions is absorbed in the world's oceans. As a result, ocean environments are experiencing a double insult. The oceans are simultaneously becoming warmer and more acidic. Both these changes have unfavorable consequences for ocean species and the humans who depend upon them for food and their livelihoods.

22. The one carbon-free electricity and process-heat-generating technology that could help diminish global warming is nuclear power, but public ambivalence or outright hostility and a lack of political leadership will preclude this technology from making a significant contribution to solving the problem until well into the second half of this century.

23. The single most important action that could significantly slow global warming would be the implementation of a substantial, enforceable, global carbon tax. A carbon tax could tip the

balance in favor of every technology and policy to reduce carbon emissions.

24. Ice is a vital buffer for the earth's climate. As the planet warms, the earth's "permanent" ice masses are melting away in the geological blink of an eye. The loss of this buffering mass will result in more dramatic annual temperature swings and significantly altered weather patterns.

25. Dozens of strategies are being pursued to reduce CO_2 emissions. Great strides will continue to be made, and per capita CO_2 emissions will begin to decrease. These are good developments. Everything that can be done to drive down emissions will lessen the severity and delay the onset of the worst consequences of global warming. However, we're not doing enough and we're not doing enough quickly enough for future generations to escape unpleasant changes to "life as we've known it."

26. The countermeasures needed to effectively "beat" global warming would be so expensive and so disruptive that the public would reject them. We're faced with a "Catch 22" – mutually conflicting requirements to resolve a dilemma.

27. Because we are not going to "beat" global warming, we should begin now making the changes necessary to accommodate rising sea levels, more severe weather events, droughts in some regions and flooding in others, reduced food production and warmer temperatures everywhere. The sooner we start making those changes, the more gradual and less onerous they'll be.

The Definition of Insanity . . .

By one definition, "insanity" is doing the same thing over and over and expecting a different result. If there's an element of truth in this axiom, then I guess most of us qualify as insane from time to time. We all sometimes continue unproductive behaviors long after our rational selves are painfully aware of the disconnect between what we want or need and what we're doing. We all hang on too long to opinions that defy reason or fly in the face evidence to the contrary.

In this book, I've explained why doing what we have been doing isn't going to "beat" global warming and why we can't and won't do what we need to. I'm hopeful that eventually we'll realize that we can't stay on this path and get the different outcome we all want. I'm hopeful that eventually we'll learn to think differently and embrace at least some of the strategies described here to limit the damage we're doing to our only home.

"Knowledge is power," denial is not a strategy and complacency is not an option.

Your Turn

That's my assessment. What's yours?

If you still think we can "beat" global warming, tell me what I've gotten wrong at BlueOasisNoMore.com. Tell me how we're going to squeeze fossil fuels out of the world's energy economy and restore the earth's fragile energy balance.

Blue Oasis No More: The Haiku

**Earth, source of all life,
Your fragile balance undone.
We did this. Our bad.**

The End(?)

EPILOGUE

"Turn your wounds into wisdom."
Oprah Winfrey

Imagine you're the coach of a high school basketball team. It's the homecoming game. It's halftime. Your team is losing by a score of 54 – 17. It would be an understatement to say it's not going well.

On the way back to the locker room, your star player twisted an ankle. Half your team's fans have left to get pizza and the other half is still in the stands, but they're otherwise absorbed on their phones. Short of divine intervention, it's pretty clear that the game's not going to end in a glorious comeback victory. And besides, you know that banking on divine intervention has never been a high-percentage strategy for success.

The door to the locker room closes. You address the team.

"OK team, I know things look pretty grim, but . . ."

I feel like that coach.

I've never coached an athletic team, but I think my halftime "pep" talk would concede that we weren't going to win the game. I'd probably try to appeal to my team's sense of pride. I'd try to convince them that losing the second half of the game by fewer points than the first would be their victory. As lame as that sounds, that's the best I've been able to come up with. It's probably a good thing I was never a coach.

In the case of global warming, there's much more on the line than "team pride."

We're not going to win the global warming "game," but by doing everything we can as quickly as we can, and by embracing solutions we may have previously rejected, we may be able to lose the second half of the game by fewer points than we lost the first.

APPENDIX A

"QUOTABLE QUOTES"

"I heard it through the grapevine."

Lyrics by Barrett Strong
and Norman Whitfield, 1966

"Global warming is the ultimate form of globalization."

"Nature is the perfect 'integrator.' It takes all its inputs, big and small, processes them, and churns out a response. That outcome is sometimes to our liking, sometimes not, but 'nature' (the physical world) doesn't care how we feel."

"If all of humanity, every single human being on Earth, were to lay down (like logs at a lumber mill) in New York City's Central Park, that stack of humanity would reach just 1,308 feet into the sky – more than 450 feet shorter than the Freedom Tower. How could such a small volume of arguably "intelligent" living matter have such a large impact on an entire planet, know that it was destroying its only home and still do almost nothing about it?"

"Forget about asteroids crashing into the earth. We're on a collision course with ourselves and we're not making the mid-course corrections necessary to avert a disaster."

"Too many people who genuinely care about global warming think that all we have to do is install solar panels on every rooftop and park an EV in every garage. They're wrong."

"People don't like to change, and they especially don't like being forced to change."

"Wind turbines and solar panels are carbon-free, proven, and inexhaustible. Unfortunately, the 'Achilles heel' of both technologies is that they're intermittent resources."

"Combating global warming will be like having a root canal that lasts for decades so that our grandchildren will feel less pain."

"As reserves of natural resources (fresh water, topsoil, fish populations, biodiversity, rare minerals, land ice, etc.) are drawn down at unsustainable rates, we'll have to expend more energy, not less, to fulfill even the basic needs of food, water and shelter for the world's growing population."

"Public support for measures to combat climate change will ebb and flow, but don't hold your breath waiting for Americans to voluntarily open their wallets to save a future they'll never personally experience."

"New technologies are seldom quite as cheap, reliable, and effective as they appear in the laboratory. The real world has a way of taking the luster off most bright shiny objects."

"The logistical accomplishment of processing and delivering four billion gallons of refined petroleum products every day would be something to celebrate in amazement . . . if it weren't KILLING THE PLANET."

"Combating climate change is not only an issue of inter-generational justice, it's also a quality-of-life and survival imperative."

"Unkept promises of a cheap and easy transition away from fossil fuels could doom the process. It's better for the public to understand what's coming than be surprised, resentful and even hostile when a more expensive energy future unfolds."

APPENDIX B

ABOUT THE AUTHOR

**"Those who dance are considered quite mad
by those who can't hear the music."**

Anne Louise Germaine de Staël, 1813

This book isn't about me, but since I've never appeared in print before and because I'm attempting to plant some potentially controversial ideas in your head, you have a right to know a little something about my background, my values **and my biases**.

I'm an optimist. Making this claim may seem odd, considering the message I'm delivering in this book. On the issue of global warming, my inner optimist concedes to my inner realist. And just to be clear, I wish that the conclusions I've reached in this book **were not so**. I would be delighted to discover that there's some fatal flaw in my argument or that there's some **safe, proven, technological "silver bullet"** that will protect us from the ravages of global warming. There isn't.

I grew up in a household where "Waste Not, Want Not" could have been embroidered, framed, and hung on the wall where any "normal" household would have displayed "Home Sweet Home." My dad was a Scotsman who proudly embraced the frugal stereotype of his homeland. However, I believe the Great Depression, rather than his heritage, did more to shape the values of both him and my mother. Growing up, the refrain heard most often from my father was "Turn out the lights." or "Turn off the water." He was ahead of this time. Mom was nearly as frugal. God bless her for often eating, rather than discarding, the cookies she mistakenly left in the oven a little too long. In many ways, I grew up a "child of the Great Depression," not having lived through it of course, but having the values it instilled in my parents so deeply imprinted on me that I might as well have.

My public high school (Bethesda-Chevy Chase HS, 1965) was top-notch, and I had the further good fortune of attending some great universities (Cornell University, BS, Engineering Physics, 1969; Massachusetts Institute of Technology, MS, NE, PhD, Nuclear Engineering, 1974; Pepperdine University, MBA, 1996).

I was graduated from Cornell in Engineering Physics. It was a rigorous program. I struggled through most of it without a clear professional direction until my senior year when I chose a few electives in nuclear engineering. I remember a visiting physics professor telling his young students that energy was the one ingredient needed to make anything and everything. That made sense to me then (still does) and the emerging civilian nuclear power industry in the late 1960s seemed like an interesting, exciting, and societally beneficial career direction.

My postgraduate program in nuclear engineering at MIT was the most gratifying academic experience of my life. The faculty and my fellow students were some of the most impressive individuals I've ever known. When graduation rolled around, doctoral degree-candidates were given the choice of being awarded a PhD (Doctor of Philosophy) or an ScD (Doctor of Science). My program was clearly in the physical sciences, but I chose to have my degree designated as a PhD because I felt my education at MIT was ultimately much more about learning how to think about a problem than it was about the diffusion of neutrons in a reactor core.

I enjoyed a variety of experiences in my career. I worked on a reactor design project at General Atomics, taught nuclear engineering at Texas A&M University, was part of a reactor physics research team at Argonne National Laboratory (West) and worked for two electric utilities (Arizona Public Service and Southern California Edison). As I moved away from engineering toward project management, I went back to school and earned an MBA at Pepperdine University.

In addition to being an engineer, I consider myself an environmentalist and a conservationist. As a young adult I had an opportunity to visit the Great Smoky Mountains National Park. A ranger there shared an expression (and a philosophy) of the early pioneers that fit comfortably with the values instilled in me as I was growing up:

"Use It Up, Wear It Out, Make It Do, Or Do Without."

The stories of settlers burning down their houses to retrieve the nails before moving to a new location really struck a chord with me. I think about that story every time an inaccurate blow from my hammer sends a nail flying.

I plant a garden annually, tomatoes mostly. When I remodeled our home a few years ago, I installed additional insulation and every energy saving feature I knew about. My wife and I both drove 2002 four-cylinder Toyota Camrys that got 30 miles to the gallon until we downsized to one car in 2020. Before the Toyotas, we had our previous cars (Volvos) even longer. I repair, or try to repair, almost everything, even things in which I know I shouldn't invest the time and effort. In short, the consumer economy in which we are all immersed would collapse in a pile of plastic and cardboard if everyone lived this way.

Since leaving college, my reading has been limited by choice to non-fiction almost exclusively. (Suggestions: <u>Sapiens: A Brief History of Humankind</u> by Yuval Noah Harari; <u>Outliers</u> by Malcolm Gladwell; <u>The Soul of America</u> by Jon Meacham; <u>The Ascent of Man</u> by Jacob Bronowski; <u>Origin Story: A Big History of Everything</u> by David Christian; <u>At Home: A Short History of Private Life</u> by Bill Bryson.)

Notably missing from this short bio is mention of any training or work experience in the atmospheric sciences. By now, however, the global warming debate no longer centers on the question: "Is it real?" Atmospheric physicists and climatologists have settled that question. The relevant question now is: "What can we do about it?" This is much more a question about transforming the world's energy economy away from fossil fuels than it is about atmospheric physics. Hopefully, my engineering experience in energy R&D, the electric utility industry, electric generation planning, my inclination to ask questions and an MBA degree give me a unique platform from which to contribute to the discussion about global warming, where we stand and what we need to do about it.

Made in United States
North Haven, CT
02 February 2023